COMBINATORIAL OPTIMIZATION II

MATHEMATICAL PROGRAMMING STUDIES

NORTH-HOLLAND PUBLISHING COMPANY – AMSTERDAM•NEW YORK•OXFORD

MATHEMATICAL PROGRAMMING STUDY 13

A PUBLICATION OF THE MATHEMATICAL PROGRAMMING SOCIETY

Combinatorial Optimization II

The Proceedings of the CO79 Conference held at the University of East Anglia, Norwhich, England 9th–12th July, 1979

Edited by V.J. RAYWARD-SMITH

T.B. Boffey
J. Clausen
J.S. Clowes
G. Cornuejols
C. McDiarmid
C.S. Edwards
L.A. Hansen
A.J.W. Hilton
A.I. Hinxman
J.-C. Picard
C.N. Potts
W. Pulleyblank
C.J. Pursglove
M. Queyranne
J.A.M. Schreuder
V.J. Rayward-Smith
L.B. Wilson
L.A. Wolsey

1980

NORTH-HOLLAND PUBLISHING COMPANY – AMSTERDAM•NEW YORK•OXFORD

This book is also available in journal format on subscription.

ISBN for this series: 0 7204 8300 X
for this volume: 0 444 86040 1

Published by:

NORTH-HOLLAND PUBLISHING COMPANY
AMSTERDAM • NEW YORK • OXFORD

Sole distributors for the U.S.A. and Canada:

Elsevier North-Holland, Inc.
52 Vanderbilt Avenue
New York, N.Y. 10017

Library of Congress Cataloging in Publication Data
Main entry under title:

Combinatorial optimization II.

(Mathematical programming study ; 13)
1. Mathematic optimization--Congresses.
2. Combinatorial analysis--Congresses. I. Rayward-Smith, V. J. II. Series.
QA402.5.C546 519 80-18710
ISBN 0-444-86040-1

PRINTED IN THE NETHERLANDS

PREFACE

In September 1977, the University of Liverpool organized the first British Conference on Combinatorial Programming. Called CP77, the conference attracted over forty delegates from both academic and industrial backgrounds and with diverse disciplines ranging from pure mathematics through engineering sciences to computing and business studies. So successful and stimulating was CP77 that it was decided to hold another conference on similar lines in 1979.

The second conference materialised as "CO79: A conference on Combinatorial Optimization" which was held at the University of East Anglia, Norwich, from 9th July to 12th July, 1979. The programme committee formed at Liverpool had decided to slightly widen the scope of the conference in the hope, thereby, of attracting more research workers from Europe and America. A call for papers was issued in September 1978 and an encouraging response guaranteed the academic success of the enterprise. In the event, twenty seven papers were presented during the conference and over sixty delegates attended from U.K., U.S.A., Canada, Europe and Israel. Much of the academic success of the conference was attributable to the stimulating presence of our invited guests: Ailsa Land (London School of Economics & Political Science), Nicos Christofides (Imperial College, London), Michael Dempster (Oxford), Eugene Lawler (Berkeley, U.S.A.), Lesley Valiant (Edinburgh) and Laurence Wolsey (Louvain-la-Neuve, Belgium).

This proceedings consists of just a selection of the papers presented at the conference. The programme committee wishes to thank all the contributors for their work and the referees for their help in the difficult task of selecting the papers to be included in this proceedings. These papers reflect the major themes of the conference which included NP—hard problems, the design and analysis of heuristics and complexity theory. Contributions range from results in pure mathematics to applications to very practical problems. The multi-disiplinary nature of the subject makes for an exciting conference and it is hoped that it will become a biennial event in Britain. The next conference in the series is planned to be at the University of Stirling, Scotland.

V.J. Rayward-Smith
Conference Coordinator

CONTENTS

CO79: COMMITTEE MEMBERS

Programme Committee: Dr. B. Boffey (Liverpool), Dr. B. Carré (Southampton), Dr. F. Dunstan (University College, Cardiff), Mr. C. Edwards (Birmingham), Dr. G. Mitra (Brunel), Dr. Susan Powell (London School of Economics & Political Science), Dr. V.J. Rayward-Smith (East Anglia, Norwich), Dr. C. Watson-Gandy (Imperial College, London), Dr. D.J.A. Welsh (Merton College, Oxford) and Dr. L.B. Wilson (Newcastle).

Organising Committee: Dr. G.P. McKeown (East Anglia, Norwich) and Dr. V.J. Rayward-Smith (East Anglia, Norwich).

Secretary: Mrs. J. Loughlin.

Mathematical Programming Study 13 (1980) 1–7.
North-Holland Publishing Company

PERFECT TRIANGLE-FREE 2-MATCHINGS*

Gérard CORNUEJOLS

Graduate School of Industrial Administration, Carnegie-Mellon University, Pittsburgh, PA, U.S.A.

William R. PULLEYBLANK

Department of Computer Science, University of Calgary, Calgary, Alta., Canada

Received 1 February 1980

The problem of determining whether a graph has a Hamilton cycle is NP-complete whereas there exists a polynomial algorithm to determine whether a graph has a perfect 2-matching. These two problems are related to the question of determining whether a graph has a perfect triangle-free 2-matching. We give a polynomial algorithm to answer this question and to find a perfect triangle-free 2-matching if one exists.

Key words: Graph Theory, Hamiltonian Cycle, Perfect 2-matching, Triangle Cluster, Triangle-free, 2-factor.

1. Introduction

The problem of determining whether a graph $G = (V, E)$ has a *Hamilton cycle* (a simple cycle containing each node exactly once) is well-known to be NP complete (cf. Karp [4] or Aho et al. [1]). Consequently, it has been conjectured that there exists no polynomially bounded algorithm for finding a Hamilton cycle in a graph, if one exists. Indeed, unless NP = co-NP, there exists no good characterization (in the sense of Edmonds [3]) of those graphs that are not Hamiltonian.

On the other hand, there does exist a polynomial algorithm for the problem, which we call P_2, of finding a *2-factor* in a graph (a set of simple cycles containing each node exactly once) if one exists. The problem P_2 is a relaxation of the Hamilton cycle problem since every feasible solution of the latter is also a feasible solution of the former. Stronger relaxations, P_k for $3 \leq k \leq |V| - 1$, can be obtained by considering the problem of finding a 2-factor in which every cycle contains more than k edges. At present it is known that P_5 is NP-hard (Papadimitriou [5]) but the status of P_3 and P_4 is unknown.

We consider here several variations of the 2-factor problem. A *2-matching* of G is an assignment of the integers $\{0, 1, 2\}$ to the edges of G such that for each node, the sum of the integers on the incident edges is at most 2. If this sum equals 2 for every node, then we say that the 2-matching is *perfect*. We let P_1

* This work was supported in part by NSF grant ENG-7902506 and the National Research Council of Canada.

denote the problem of finding a perfect 2-matching of a graph, if one exists. P_1 is a relaxation of P_2 and moreover P_1 is known to be equivalent to the problem of finding a perfect 1-matching in a bipartite graph. The following result is well-known.

Theorem 1.1. *If G has a perfect 2-matching, then it has one for which the edges assigned the value 1 form disjoint odd cycles.*

A *triangle* in G is the edge set of a cycle with three nodes and three edges. We say that a 2-matching x is *triangle-free* if for every triangle of G there is at least one edge j for which $x_j = 0$. The main subject of interest in this paper is the problem $\bar{P}_3$: given a graph G, find a perfect triangle-free 2-matching, if one exists. Note that P_1 is a relaxation of both P_2 and $\bar{P}_3$ and, in turn, both P_2 and $\bar{P}_3$ are relaxations of P_3. However, neither $\bar{P}_3$ nor P_2 is a relaxation of the other.

In the next section we present a polynomially bounded algorithm which solves $\bar{P}_3$. Then in Section 3 we discuss several related problems.

2. The algorithm

The construction of perfect triangle-free 2-matchings makes use of the following graphical structure. A *triangle cluster* is a connected graph whose edges partition into disjoint triangles with the property that any two triangles have at most one node in common and if such a node exists, then it is a cutnode of the cluster. See Fig. 1.

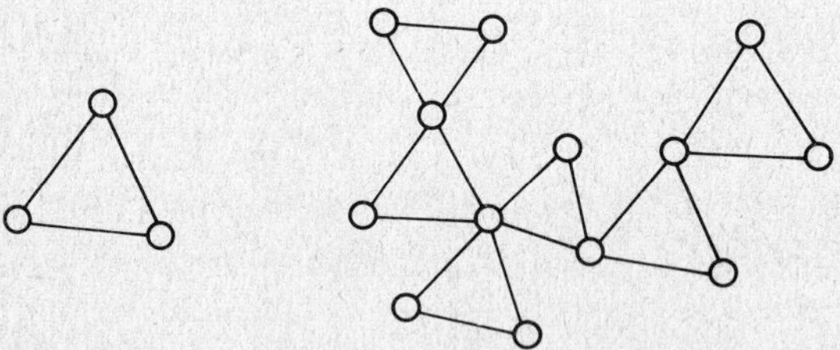

Fig. 1. Triangle clusters.

The following properties of triangle clusters are easily verified (see Cornuéjols and Pulleyblank [2] for details).

Proposition 2.1. *A triangle cluster with k triangles has* $2k+1$ *nodes.*

Proposition 2.2. *A triangle cluster does not have a perfect triangle-free 2-matching.*

Proposition 2.3. *If any node is deleted from a triangle cluster, then the resulting graph has a unique perfect triangle-free 2-matching.*

Proposition 2.4. *There is a unique maximum length simple path joining any two nodes of a triangle cluster. This path has even length and moreover if all the nodes in this path are deleted, then the resulting graph has a unique perfect triangle-free 2-matching.*

It is straightforward to develop procedures for constructing the triangle-free 2-matchings of Propositions 2.3 and 2.4. Moreover it can be easily shown using Theorem 2.5, proved at the end of this section, that a triangle cluster is the only graph that satisfies the following properties:

(i) it does not have a perfect triangle-free 2-matching,

(ii) if any node is deleted, then the resulting graph does have a perfect triangle-free 2-matching.

In the course of the algorithm we grow a certain type of alternating tree. The nodes of the alternating tree F may be of two types. A *real node* of F is simply a node of G. A *cluster node* of F is a triangle cluster contained in G. The edges of F are edges of G where we consider an edge j to be incident with a cluster node of F if j is not in the triangle cluster but one end of j is a node of the cluster. See Fig. 2.

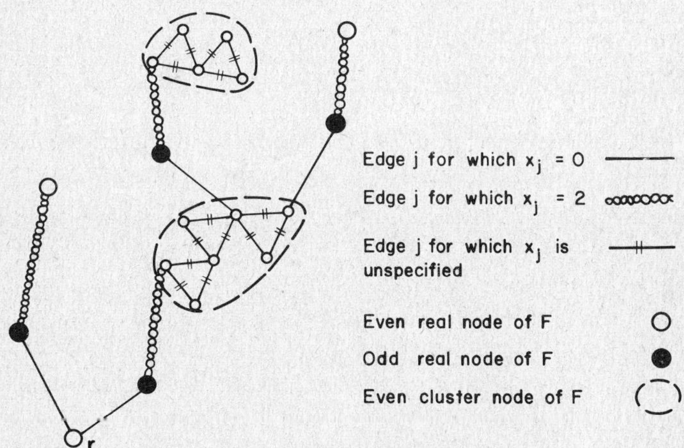

Fig. 2. Alternating tree.

The tree F is *rooted* at a node r (which may be a real node or a cluster node). The nodes of F are designated as being *odd* (*even*) if the number of edges of F in the path to r in F is odd (even). Odd nodes of F will always be real nodes.

An alternating tree is always defined relative to a triangle-free 2-matching x (which will not be perfect) and must satisfy the following conditions.

(i) In every path in F from r to another node of F the values x_j for the edges j in the path are alternately 0 or 2.

(ii) Each odd node of F is incident with exactly two edges of F.

(iii) For every edge j which is not an edge of F but is incident with a node of F we have $x_j = 0$.

Note that the conditions (i)–(iii) do not impose any restrictions on the value of x_j when j belongs to a cluster node of F. However, in view of Proposition 2.3, we know that x_j will be uniquely defined for each edge j of a cluster node.

The algorithm starts with a (not necessarily perfect) triangle-free 2-matching x, which may be defined by letting $x_j = 0$ for all $j \in E$. It will then attempt to "improve" x, if possible, in the following way. If there is a node r such that $x_j = 0$ for every j incident with r, then the algorithm grows an alternating tree rooted at r. This tree growth continues until either a means of augmenting the matching is discovered or no further growth is possible. In this latter case the algorithm discovers a structure which shows that no perfect triangle-free 2-matching exists.

We now describe the algorithm in detail.

Step 0 [Initialization]: Let x be any triangle-free 2-matching of G for which the edges assigned the value 1 form disjoint odd cycles. (For example $x_j = 0$ for all $j \in E$.)

Step 1 [Optimality Test]: If x is perfect, then terminate. Otherwise find a node r such that $x_j = 0$ for every edge j incident with r. We now begin growing an alternating tree F rooted at r. Initially, F consists of a single even node, namely r.

Step 2 [Edge Selection]: Find, if one exists, an edge j joining an even node u of F to a node v which is not an odd node of F. If no such edge exists, terminate, as no perfect triangle-free 2-matching exists (a consequence of Proposition 2.2). If such an edge is found, then there are four cases.

Case 1: v is not a node of F and is incident with an edge k for which $x_k = 2$. Go to Step 3 where we grow the tree.

Case 2: v is not a node of F and $x_k = 0$ for every edge k incident with v. Go to Step 4 where we augment the matching.

Case 3: v is not a node of F and there are two edges h and l incident with v for which $x_h = x_l = 1$. Go to Step 5 where we augment.

Case 4: v is an even node of F. Go to Step 6 where we augment or cluster.

Step 3 [Tree Growth]: Let w be the node incident with k which is different from v. Grow F by adjoining edges j and k and nodes v and w. Thus v becomes an odd node of F and w becomes an even node of F. Go to Step 2.

Step 4 [Simple Augmentation]: Set $x_j = 2$. Then traverse the path in F from u to the root r alternately lowering and raising by 2 the value x_h for each edge h of F encountered in this path. After this change any cluster node K of F will have exactly one real node $w \in K$ incident with an edge k of F for which $x_k = 2$. As a consequence of Proposition 2.3, it is simple to correct the values of the edges of the cluster so that x will be a triangle-free 2-matching that perfectly matches every real node of K. We now "throw away" F and any clusters formed and go to Step 1.

Step 5 [Cycle Breaking Augmentation]: h and l belong to an odd cycle P of G such that $x_k = 1$ for every edge k in P. We now travel around P starting with h setting $x_k = 0$ or 2 alternately for each edge k until we reach edge l. Then x_h and x_l will both be 0. Consequently, every edge incident with v is assigned the value 0; so we go to Step 4 (after which we return to Step 1).

Step 6 [Augment or Cluster]: Edge j added to F creates an odd cycle P' (which may be a loop if $u = v$). Edge j together with edges of F and edges in clusters forms a unique maximum length odd polygon P in G by Proposition 2.4. (We define a *polygon* as the edge set of a cycle.) If P is a triangle, then we go to Step 6a where we cluster. If P is not a triangle, we go to Step 6b where we augment.

Step 6a [Cluster] (Fig. 3.): Create a new triangle cluster C containing P and the clusters making up its nodes, if any of them are not real nodes of G. Now C is an even node of F. Go to Step 2.

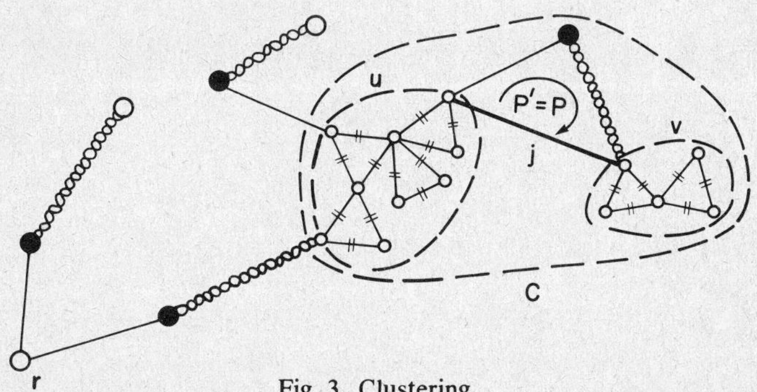

Fig. 3. Clustering.

Step 6b [Augment] (Fig. 4.): We have an odd cycle P' joined by an even length path π from P' to the root r. Set $x_j = 1$ for every edge in P' and alternately set the edges of π to 0 or 2 until r is reached. If π and P' contain any cluster

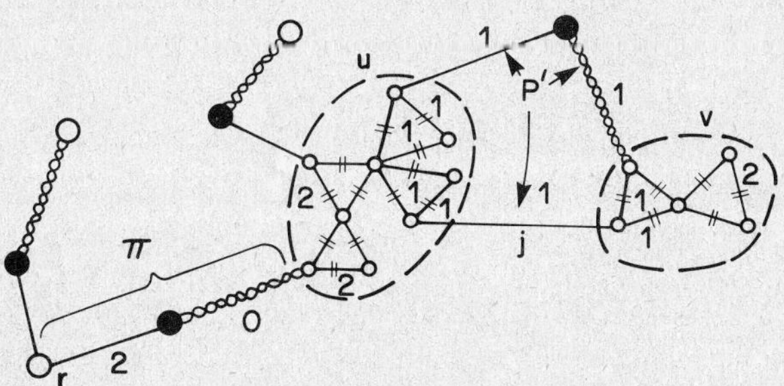

Fig. 4. Augmentation.

nodes, the matching in these cluster nodes can be modified by Propositions 2.3 and 2.4, respectively. "Throw away" F and any clusters which have been formed and go to Step 1.

Remarks on the algorithm. (i) It is straightforward to show that an upper bound on the running time of the algorithm is $O(|V|^3)$.

(ii) If the algorithm terminates in Step 1, then clearly a perfect triangle-free 2-matching has been found. Now suppose that no perfect triangle-free 2-matching exists. Then the algorithm terminates in Step 2. Let X be the set of odd nodes of F. Since every edge incident with an even node of F has as its other end an odd node of F, it follows that each even node of F will be a connected component of $G[V-X]$ (the subgraph of G induced by $V-X$). Moreover F has $|X|+1$ even nodes. Conversely, in view of Proposition 2.2, it is clear that if there exists a set $X \subseteq V$ such that more than $|X|$ components of $G[V-X]$ are triangle cluters on isolated nodes, then no perfect triangle-free 2-matching of G can exist. Thus we have the following.

Theorem 2.5. *The graph $G=(V, E)$ has a perfect triangle-free 2-matching if and only if for every $X \subseteq V$ the graph $G[V-X]$ has at most $|X|$ components which are triangle clusters or isolated nodes.*

It is interesting to compare this result with the characterization of those graphs that have perfect 2-matchings.

Theorem 2.6. (Tutte [6]). *G has a perfect 2-matching if and only if for every $X \subseteq V$ the graph $G[V-X]$ has at most $|X|$ isolated nodes.*

3. Related problems

Cornuéjols and Pulleyblank [2] consider a weighted version of the problem studied here. Let $c=(c_j: j \in E)$ be a vector of arbitrary edge weights. The weighted problem is to find a (not necessarily perfect) triangle-free 2-matching x which maximizes $\sum (c_j x_j: j \in E)$. We describe a polynomial algorithm for this problem and prove the following polyhedral characterization theorem[1].

Theorem 3.1. *The convex hull of the set of triangle-free 2-matchings of a graph G is the solution set of the following linear system:*

$$x_j \geq 0 \quad \textit{for all } j \in E,$$

$$\sum (x_j: j \textit{ incident with } i) \leq 2 \quad \textit{for all } i \in V,$$

[1] *Added in proof*: This result has been independently obtained by J.F. Maurras.

$$\sum (x_j\colon j \in T) \leq 2 \quad \textit{for every triangle } T \textit{ of } G.$$

A variation of $\bar{P}_3$, the problem of determining whether or not G has a perfect triangle-free 2-matching is the following: Does G have a perfect 2-matching satisfying $\sum(x_j\colon j \in T) \leq 1$ for every triangle T of G. In the reference mentioned above we show that this problem is np-complete.

References

[1] A.V. Aho, J.E. Hopcroft and J.D. Ullman, *The design and analysis of computer algorithms* (Addison-Wesley, Reading, MA, 1974).
[2] G. Cornuéjols and W. Pulleyblank, "A matching problem with side conditions", *Discrete Mathematics* 29 (1980) 135–159.
[3] J. Edmonds, "Maximum matching and polyhedron with 0–1 vertices", *Journal of Research of the National Bureau of Standards* 69b (1965) 125–130.
[4] R.M. Karp, "Reducibility among combinatorial problems", in: R.E. Miller and J.W. Thatcher, eds., *Complexity of computer computations* (Plenum Press, New York) pp. 85–103.
[5] C.H. Papadimitriou, Private communication (1978).
[6] W.T. Tutte, "The factors of graphs", *Canadian Journal of Mathematics* 4 (1952) 314–328.

Mathematical Programming Study 13 (1980) 8–16.
North-Holland Publishing Company

ON THE STRUCTURE OF ALL MINIMUM CUTS IN A NETWORK AND APPLICATIONS

Jean-Claude PICARD

Ecole Polytechnique, Thiès, Sénégal

Maurice QUEYRANNE

University of Houston, Houston, TX, U.S.A.

Received 1 February 1980

This paper presents a characterization of all minimum cuts, separating a source from a sink in a network. A binary relation is associated with any maximum flow in this network, and minimum cuts are identified with closures for this relation. As a consequence, finding all minimum cuts reduces to a straightforward enumeration. Applications of this results arise in sensitivity and parametric analyses of networks, the vertex packing and maximum closure problems, in unconstrained pseudo-boolean optimization and project selection, as well as in other areas of application of minimum cuts.

Key words: Maximum Closure, Maximum Flow, Minimum Cuts, Networks, Optimization, Parametric Analysis, Sensitivity Analysis, Vertex Packing.

1. Introduction

Consider a finite directed network with positive arc capacities, and two special vertices, a source s and sink t. The problem of finding a cut separating s from t, with minimum capacity can be solved by applying any maximum flow algorithm and using the maximum-flow/minimum-cut theorem of Ford and Fulkerson. Here we consider the problem of finding *all* the minimum cuts.

It appears that this is only the problem of finding all optimum solutions to a linear programming problem. However, this is not a simple task. Consider for instance a network with n vertices and $2n-4$ arcs, namely (s, i) and (i, t) for all vertices $i \neq s$ and t, all with equal capacities (see Fig. 1): this network admits 2^{n-2} cuts separating s from t, all being minimum cuts. It follows that we cannot expect a polynomial algorithm for finding all minimum cuts.

In the next section, we show that we can associate a binary relation with every network, such that finding all minimum cuts reduces to finding all closures for this relation. There exist efficient enumerative methods for generating all closures, thus producing all minimum cuts. In addition this associated binary relation provides more insight into the structure of minimum cuts in a network.

In the last section, we mention several applications in which it is useful to know all the minimum cuts in a network or at least all the arcs which belong to

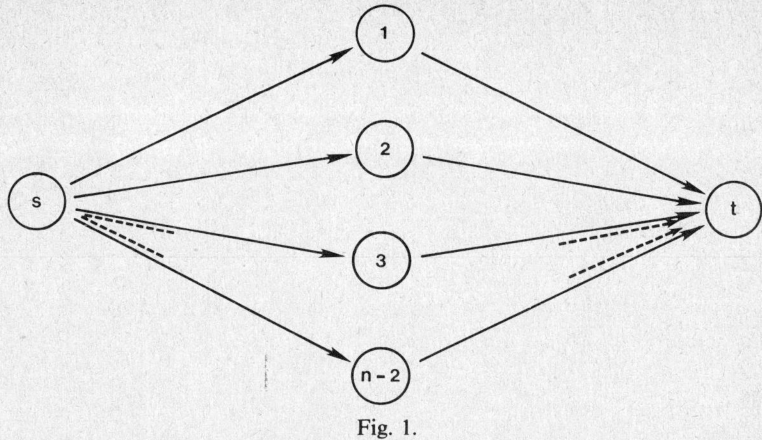

Fig. 1.

some minimum cut. In these applications, finding all the minimum cuts allows a better solution of the problem considered, or at least helps to reduce the computational burden for a subsequent algorithm.

2. Structure of minimum cuts

We are given a finite directed network $N = (V, A, c)$, with vertex set V, including a source s and a sink t, arc set A and positive arc capacities c_{ij} defined on A. Given two disjoint subsets S and T of V, we denote by (S, T) the set of all arcs in A with tail in S and head in T. When a function f is defined on A, we denote by $f(S, T)$ the sum of the values of f on the arcs in (S, T). A *cut* separating s from t is any arc set $(S, \bar{S})$ where $s \in S$, $\bar{S} = V - S$ is the complement of S and $t \in \bar{S}$. The *capacity of a cut* $(S, \bar{S})$ is $c(S, \bar{S})$, the sum of the capacities of the arcs in $(S, \bar{S})$. By a *minimum cut* we mean a cut separating s from t with minimum capacity.

Given a binary relation R defined on V, whenever iRj we say that i is a *predecessor* of j and j is a *successor* of i. A subset $C \subseteq V$ is a *closure* [17] for R iff for all vertices $i, j \in V$, the conditions $i \in C$ and iRj imply $j \in C$. (This is sometimes called a *hereditary* subset for R (see [6])).

Consider any maximum flow f in N. From the maximum-flow/minimum-cut theorem of Ford and Fulkerson [4], we know that such a flow exists and has a value equal to the minimum capacity of a cut. We assume that such a maximum flow is given, since it can be computed by efficient algorithms.

Theorem 1. *Let f be any maximum flow in N. Define a relation R on the vertex set V as follows:*

$$iRj \quad \textit{iff} \ ((i, j) \in A \ \textit{and} \ f_{ij} < c_{ij}) \quad \textit{or} \quad ((j, i) \in A \ \textit{and} \ f_{ji} > 0). \tag{1}$$

Then a cut $(S, \bar{S})$ separating s from t is a minimum cut if and only if S is a closure for R containing s and not t.

Proof. Consider a cut $(S, \bar{S})$ separating s from t. For any feasible flow f in N, we have

$$c(S, \bar{S}) \geq f(S, \bar{S}) - f(\bar{S}, S) \tag{2}$$

and equality holds if and only if both f is a maximum flow and $(S, \bar{S})$ is a minimum cut. Then for all arcs $(i, j) \in (S, \bar{S})$ we have $f_{ij} = c_{ij}$ and for all arcs (j, i) with $i \in S$ and $j \in \bar{S}$ we have $f_{ji} = 0$. This implies that S is a closure for R, containing s and not t, for otherwise there would exist two vertices i and j such that $i \in S$, $j \in \bar{S}$ and either $f_{ij} < c_{ij}$ or $f_{ji} > 0$, a contradiction. Conversely, consider a closure S for R, containing s and not t. For every arc (i, j) in $(S, \bar{S})$ we must have $f_{ij} = c_{ij}$, and for every arc (j, i) in $(\bar{S}, S)$ we must have $f_{ij} = 0$. It follows that equality holds in (2) and thus $(S, \bar{S})$ is a minimum cut.

This theorem gives more insight into the structure of minimum cuts in N. The following proposition is immediate from the definition of a closure:

Proposition 2. *Given a binary relation R on a set, if C and C' are closures for R, then $C \cup C'$ and $C \cap C'$ are also closures for R.*

Hence the following corollary [4], a proof of which requires two pages in [9]:

Corollary 3. If *$(S, \bar{S})$ and $(S', \bar{S}')$ are minimum cuts in a network N, then $(S \cup S', \overline{S \cup S'})$ and $(S \cap S', \overline{S \cap S'})$ are also minimum cuts in N.*

Given a maximum flow, the corresponding relation R can be deduced by a simple examination of all the arcs in A. Distinct maximum flows may produce different relations but the set of closures remains the same. Define the *transitive closure* $\tilde{R}$ of a binary relation R as the smallest transitive binary relation on the same set, containing R. The following proposition is easily proven:

Proposition 4. *A subset C is a closure for R in and only if it is a closure for $\tilde{R}$.*

A bit more difficult to prove is the following:

Proposition 5 (see [13]). *If R and R are transitive relations defined on the same set, such that any subset C is a closure for R if and only if it is also a closure for R', then $R = R'$.*

Thus the different binary relations defined by different maximum flows have the same transitive closure, which we call the *preorder* associated with the network N.

Consider now the problem of finding all minimum cuts in a network. After computing a maximum flow, a minimum cut is identified by the Labelling Procedure of Ford and Fulkerson [4]; this minimum cut $(S, \bar{S})$ is the one with the smallest possible source set S. Before defining the relation R, it may be useful to verify whether the minimum cut is unique: this can be performed by producing the minimum cut $(S', \bar{S}')$ with largest possible source set S', applying a "Reverse" Labelling Procedure starting from the sink (the details are left to the reader). If these two minimum cuts differ, we can define the relation R associated with the maximum flow and shrink its strongly connected components to single vertices.

The resulting relation $\bar{R}$ on the reduced set $\bar{V}$ is defined by $\bar{k}\bar{R}\bar{l}$ iff iRj for some $i \in \bar{k}$ and $j \in \bar{l}$; it is acyclic, that is a precedence relation (or a partial order). After eliminating the component T containing the sink t, and all its predecessors (which cannot belong to a closure not containing T) and the component S containing the source s, and all its successors (which must belong to a closure containing S) we are left with a further reduced relation, every closure of which induces (after addition of S and all its successors) a minimum cut in N.

For enumerating all these closures, we can apply procedures of Gutjahr and Nemhauser [7], Schrage and Baker [25] or Lawler [11]. These last two procedures appear very efficient, requiring very little bookkeeping effort for every closure generation.

Example. Consider the network given by Fig. 2. A maximum flow is given in Fig. 3. The associated relation R appears on Fig. 4, where an arc (i, j) represents iRj and a bidirected arc (i, j) stands for both iRj and jRi (when the corresponding arc has flow strictly between zero and its capacity). The strongly connected components are $S = \{s, 2\}$, $T = \{t, 8, 12\}$, $V1 = \{1\}$, $V3 = \{3, 7\}$, $V4 = \{4\}$, $V5 = \{5, 9\}$ and $V6 = \{6, 10, 11\}$, and after shrinking these to a single vertex, the resulting relation $\bar{R}$ is given by Fig. 5. Here $V3$ is a successor of S and $V6$ is a predecessor of T. The other components $V1$, $V4$ and $V5$ are all predecessors of S and successors of T, and they induce the relation given in Fig. 6. This relation admits six closures C, each one defining a minimum cut $(X, \bar{X})$, as follows:

$$
\begin{aligned}
&C = \emptyset && \text{and} \quad X = S \cup V3,\\
&C = \{V1\} && \text{and} \quad X = S \cup V3 \cup V1,\\
&C = \{V1, V4\} && \text{and} \quad X = S \cup V3 \cup V1 \cup V4,\\
&C = \{V1, V4, V5\} && \text{and} \quad X = S \cup V3 \cup V1 \cup V4 \cup V5,\\
&C = \{V1, V5\} && \text{and} \quad X = S \cup V3 \cup V1 \cup V5\\
&C = \{V4\} && \text{and} \quad X = S \cup V3 \cup V4.
\end{aligned}
$$

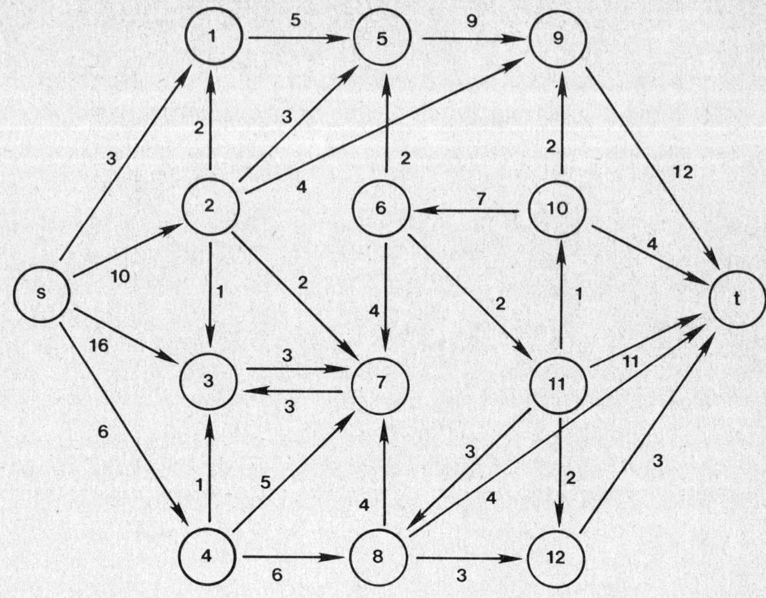
1
5
5
9
9
3
2
3
2
4
6
2
7
10
2
12
s
10
1
2
4
2
1
4
t
16
3
7
11
11
3
3
6
3
2
3
1
5
4
4
4
8
12
6
3

Fig. 2.

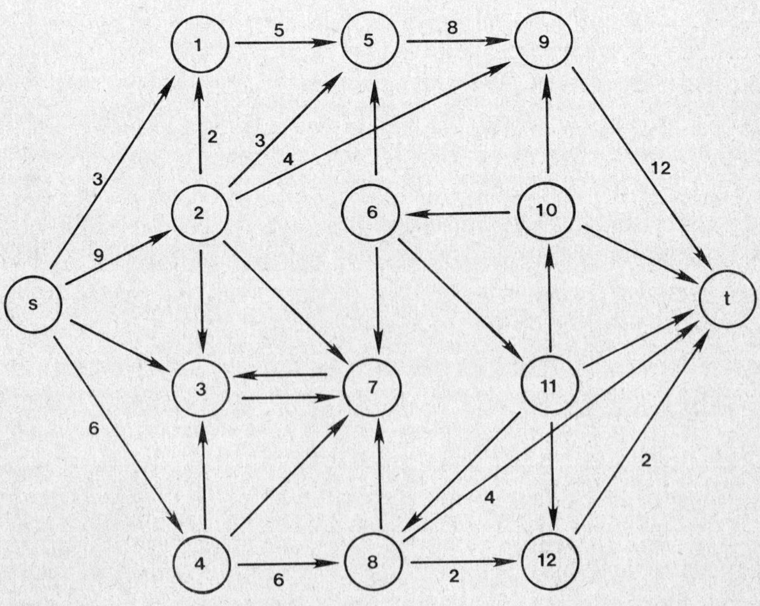
1
5
5
8
9
2
3
4
3
12
2
6
10
9
s
t
3
7
11
6
2
4
4
8
12
6
2

Fig. 3.

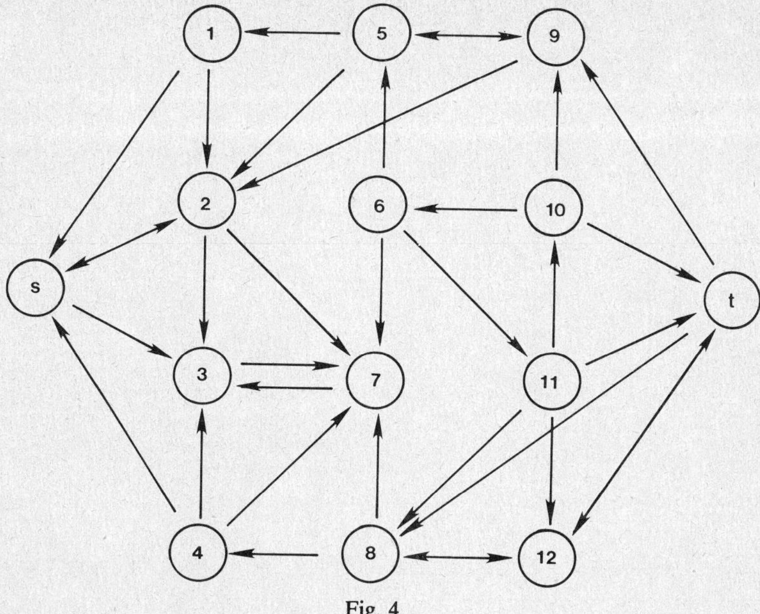

Fig. 4.

3. Application and extensions

The main result of the previous section provides more insight into the structure of minimum cuts in a network. In this section we mention several domains of applications for this result.

The structure revealed by the preorder associated with the network can be used to simplify sensitivity and parametric analyses of the maximum flow. In sensitivity analysis, it is required to find all the arcs such that a modification (increase or decrease) of the capacity of one of them implies a modification of the maximum value of a flow. It is clear that only saturated arcs are to be considered, and that any reduction in the capacity of an arc which belongs to some minimum cut implies a reduction in the flow value. These arcs are identified as follows:

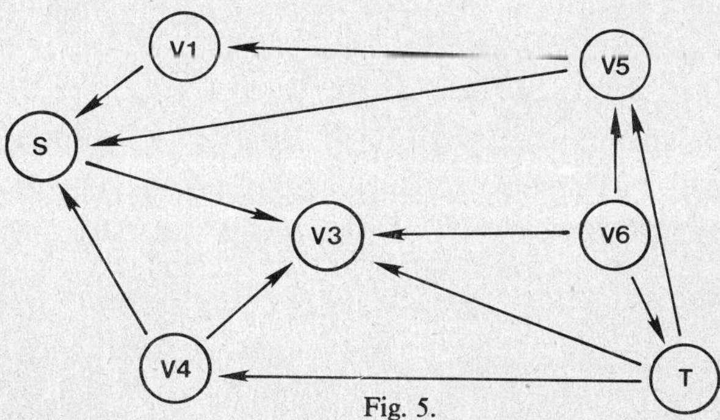

Fig. 5.

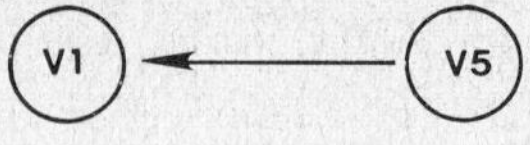

Fig. 6.

Corollary 6. *A saturated arc belongs to some minimum cut if and only if its ends do not lie in the same strongly connected component of the relation R.*

On the other hand, an increase in the capacity of an arc allows an increase in the flow value if and only if this arc has its tail in the strongly connected component containing the source (or some successor of it) and its head in the component containing the sink (or some predecessor of it). Similar results apply to various parametric analyses such as adding new arcs or nodes [27], finding the most vital arcs [12, 26, 28] or nodes [8] and in the analysis of dynamic maximum flow [14, pp. 128–151]. One practical application of dynamic maximum flow is the modeling of building evacuation [5]: given the minimum evacuation time, it is desired to detect all evacuation bottlenecks which may cause delays and to which special attention must be given; these are precisely the arcs which belong to some minimum cut.

The number of minimum cuts can be obtained as a by-product of their enumeration. Consider a communication network $N=(V,A)$ subject to arc failures, and assume that every arc has a probability p of failure and that all the failures are independent events. If we denote by A_k the number of subsets of k arcs in A which disconnect s from t, then this probability of disconnection is

$$P(s,t)=\sum_{k=1}^{m} A_k p^k (1-p)^{|A|-k}$$

see [3, pp. 432–434]. If we consider very reliable networks, we are interested in cases where p is very small and a good approximation for $P(s,t)$ is $A_{k^*}p^{k^*}(1-p)^{|A|-k^*}$, where k^* is the minimum number of arcs in a cut separating s from t, and A_{k^*} is the number of these minimum cuts.

Identifying all minimum cuts is also useful whenever a problem is reduced to finding a minimum cut in a network satisfying additional constraints. Consider for example the vertex packing problem in a vertex-weighted undirected graph [15]: solving a linear programming relaxation of one integer programming formulation can be achieved by finding a minimum cut in a related bipartite network, producing a solution with values 0, 1 or $\frac{1}{2}$ and it is desired to find a solution with the maximum number of 0, 1 components [19]; this can be achieved by classical sensitivity analysis [15], or by a specialized algorithm [18] and also by identifying all minimum cuts and retaining the one producing the most integral solution.

Another problem amenable to a minimum cut solution, which has significant practical implications is the maximum closure problem [17], a generalization of the selection problem [2, 24]. In investment application, or in mining engineering, it is desirable to obtain all solutions with maximum weight, from which a "best" one is selected on the basis of ill-formulated constraints or objectives (e.g. [13]). In mathematical programming, the unconstrained maximization (or minimization) of a pseudo-boolean polynomial can be approached by solving a related maximum closure problem [20]; the corresponding solution may be overestimated, by omission of some nonlinear terms with negative costs which cannot be covered by other positive terms (see [20]) for further details) and identification of all optimal closures may be useful by producing several tentative solutions from which the best one can be retained as an incumbent in a subsequent branch-and-bound algorithm. There are several other applications of minimum cuts and maximum closures, which may benefit from identification of all optimal solutions and the reader is refered to [21] for a more detailed survey.

The results of this paper can be extended to undirected networks and to networks with lower capacities. Any undirected network can be converted to a directed network by arbitrarily directing its edges and adding some source and sink-arcs, such that the relative capacities of the cuts remain unchanged [22]. Hence all the minimum cuts of an undirected network can be found after this reduction by applying the previous results. Among possible applications are a layout problem of electrical connexions on a line [1] and the design of optimum communication networks [10]. The results of this paper also extend to networks with lower capacities [4], and this is left to the reader as an exercise. The project time/cost tradeoff problem of critical path analysis can be approached by finding minimum cuts in the project network, which includes both lower and upper capacities [16]. The authors note that the minimum cut is not necessary unique and state: "The practical significance of this fact is that a decision based on other than cost must be rendered to select a minimal cut set" [16, p. 396]. Clearly, this selection process is best performed when all minimum cuts have been identified.

References

[1] D. Adolphson and T.C. Hu, "Optimal linear ordering", *Society for Industrial and Applied Mathematics Journal of Applied Mathematics* 25 (1973) 403–423.

[2] M.L. Balinski, "On a selection problem", *Management Science* 17 (1970) 230–231.

[3] D.W. Davies and D.L.A. Barber, *Communication networks for computers* (Wiley, Chichester, Great Britain, 1973).

[4] L.R. Ford and D.R. Fulkerson, *Flows in networks* (Princeton University Press, Princeton, NJ, 1962).

[5] R.L. Francis and P.B. Saunders, "EVACNET: Prototype network optimization models for building evacuation", Report NBSIR 79-1738, National Bureau of Standards, Washington, DC (1979).

[6] G. Gratzer, *Lattice Theory: first concepts and distributive lattices* (W.M. Freeman and Co., San Francisco, CA., 1971).

[7] A.L. Gutjahr and G.J. Nemhauser, "An algorithm for the line balancing problem", *Management Science* 11 (1964) 308–315.

[8] Han Chang, "Funding the *n* most vital nodes in a flow network", Dissertation, University of Texas at Arlington, TX (1972).

[9] T.C. Hu, *Integer programming and network flows* (Addison-Wesley, Reading, MA, 1970).

[10] T.C. Hu, "Optimum communication spanning trees", *Society for Industrial and Applied Mathematics Journal of Computing* 3 (1974) 188–195.

[11] E.L. Lawler, "Efficient implementation of dynamic programming algorithms for sequencing problems", Report bw 106/79, Stitchting Mathematisch Centrum, Amsterdam, The Netherlands (1979).

[12] S.M. Lubore, H.D. Ratliff and G.T. Sicilia, "Determining the most vital link in a flow network", *Naval Research Logistic Quarterly* 18 (1971) 497–502.

[13] L.F. McGinnis and H.L.W. Nuttle, "The project coordinators' problem", *Omega* 6 (1978) 325–330.

[14] E. Minieka, *Optimization algorithms for networks and graphs* (Marcel Dekker Inc., New York, 1978).

[15] G.L. Nemhauser and L.E. Trotter, "Vertex packings: structural properties and algorithms", *Mathematical Programming* 8 (1975) 232–248.

[16] S. Phillips Jr. and M.E. Dessouky, "Solving the project time/cost tradeoff problem using the minimal cut concept", *Management Science* 24 (1977) 393–400.

[17] J.-C. Picard, "Maximal closure of a graph and application to combinatorial problems", *Management Science* 22 (1976) 1268–1272.

[18] J.-C. Picard and M. Queyranne, "Vertex packings: (VLP)—reductions through alternate labeling", Technical report EP75-R-47, Ecole Polytechnique de Montréal, Que., Canada (1975).

[19] J.-C. Picard and M. Queyranne, "On the integer-valued variables in the linear vertex packing problem", *Mathematical Programming* 12 (1977) 97–101.

[20] J.-C. Picard and M. Queyranne, "Networks graphs and some nonlinear 0–1 programming problems", Technical report EP77-R-32, Ecole Polytechnique de Montréal, Que., Canada (1977).

[21] J.-C. Picard and M. Queyranne, "Selected applications of the maximum flow and minimum cut problems", Tech. Rept. EP79-R-35, Ecole Polytechnique de Montréal, Montréal, Qué., Canada (1979).

[22] J.-C. Picard and H.D. Ratliff, "Minimum cuts and related problems", *Networks* 5 (1975) 357–370.

[23] M. Queyranne, "Anneaux achevés d'ensembles et préordres", Technical report EP77-R-14, Ecole de Montréal, Que., Canada (1977).

[24] J.M.W. Rhys, "A selection problem of shared fixed costs and network flows", *Management Science* 17 (1970) 200–207.

[25] L. Schrage and K.R. Baker, "Dynamic programming solution of sequencing problems with precedence constraints", *Operations Research* 26 (1978) 444–449.

[26] G.T. Sicilia, "Finding the *n* most vital links in a network", Dissertation, University of Florida, Gainesville, FL, (1970).

[27] D.M. Topkis, "Monotone minimum node-cuts in capacitated networks", Research report ORC 70-39, University of California, Berkeley, CA, (1970).

[28] R. Wollmer, "Sensitivity analysis in networks", Technical report ORC 65-8, University of California, Berkeley, CA, (1965).

Mathematical Programming Study 13 (1980) 17–25.
North-Holland Publishing Company

CLUTTER PERCOLATION AND RANDOM GRAPHS

COLIN McDIARMID*
London School of Economics, Houghton Street, London, Great Britain

Received 1 February 1980

The "clutter percolation theorem" is presented and from it are deduced various results on paths in random graphs and digraphs.

Key words: Bethe Tree, Clutter, Connectedness, Graphs, Hamiltonian Cycles, Paths, Percolation, Probability, Random.

1. Introduction

I introduce here a general theorem on "clutter percolation" and deduce from it various qualitative results concerning paths and connectedness in random graphs and digraphs. More general results (and a proof of the clutter percolation theorem) may be found in [11]. An example of the sort of result we obtain here is the following.

Let V be a set of n (≥ 2) vertices and let $0 < p < 1$. We use $G_{n,p}$ to denote the random graph on V in which the $\frac{1}{2}n(n-1)$ possible edges occur independently with probability p. Similarly, we use $D_{n,p}$ to denote the random digraph on V in which the $n(n-1)$ possible edges occur independently with probability p. Then the probability that $G_{n,p}$ is Hamiltonian is less than the probability that $D_{n,p}$ is Hamiltonian. (Recall that a graph or digraph is Hamiltonian if it contains a closed path or cycle going through each vertex exactly once.)

2. Clutter percolation

In this section I introduce the idea of clutter percolation (following Oxley and Welsh [12]) and state (without proof) one general theorem.

Let I be a finite non-empty set and let $0 \leq p \leq 1$. We suppose that each element i of I is independently *open* with probability p and *closed* with probability 1-p. A subset J of I is *open* if each of its elements is open. Now let $\mathscr{C}$ be any collection of subsets of I. The *percolation probability* (or *reliability*) $P(\mathscr{C}, p)$ is the probability that some set in $\mathscr{C}$ is open.

A *clutter* (or Sperner family) on I is a collection of pairwise incomparable

*This research was supported in part by Canadian NRC grant A9211.

subsets of I. If $\mathscr{A}$ is any collection of subsets of I and $\mathscr{C}$ is the clutter of minimal members of $\mathscr{A}$, then of course $P(\mathscr{C}, p) = P(\mathscr{A}, p)$. Thus we restrict our attention here to clutters.

Now let $\mathscr{C}$ be a clutter on I and let $\sim$ be an equivalence relation on I. We need to consider two (dual) ways in which $\mathscr{C}$ and $\sim$ may be related. The following two conditions (C) and (C*) feature throughout this paper.

(C) $\quad a \sim b, \quad a \neq b, \quad C \in \mathscr{C} \;\Rightarrow\; \{a, b\} \not\subseteq C,$

(C*) $\quad a \sim b, \quad A, B \in \mathscr{C}, \quad a \in A \setminus B, \quad b \in B \setminus A$
$\quad \Rightarrow \; \exists C \in \mathscr{C}, \quad C \subseteq (A \cup B) \setminus \{a, b\}.$

For example let I be the edge set of a digraph D without parallel edges, and let $\sim$ be the equivalence relation which makes opposite edges equivalent (so that $(u, v) \sim (v, u)$). If $\mathscr{C}_1$ is the clutter of edge sets of Hamiltonian cycles in D, then $(\mathscr{C}_1, \sim)$ satisfies condition (C). Now let s and t be specified vertices in D and let $\mathscr{C}_2$ be the clutter of minimal edge sets of paths from s to t. Then $(\mathscr{C}_2, \sim)$ satisfies both condition (C) and (C*)—see the first application in Section 3.

We need one more definition. Suppose that $\mathscr{C}$ and $\sim$ are as above. The *underlying clutter* $\tilde{\mathscr{C}}$ of $\mathscr{C}$ with $\sim$ is defined on the set of equivalence classes $[i]$ and is the clutter of minimal sets of the form $\{[i]\colon i \in C\}$ for C in $\mathscr{C}$.

Let us illustrate this idea. Suppose in the last example that D is the digraph associated with some underlying simple graph G; that is $D = D(G)$ is obtained from G by replacing each undirected edge $\{u, v\}$ by a pair of oppositely directed edges (u, v) and (v, u). Let us identify the equivalence class $\{(u, v), (v, u)\}$ with the edge $\{u, v\}$ of G. Then $\tilde{\mathscr{C}}_1$ is the clutter of edge sets of Hamiltonian cycles in G (assuming that there are at least three vertices), and $\tilde{\mathscr{C}}_2$ is the clutter of minimal edge sets of paths in G from s to t.

We are now ready for the clutter percolation theorem. The proof is not difficult but it takes a little time and I spare you the details here. It is easiest and most natural in fact to prove the theorem in a more general setting (see [11]).

Before I state the theorem let us consider a small example. Let $I = \{a, b\}$, let $a \sim b$, and let $0 < p < 1$.

(a) Suppose first that $\mathscr{C}$ is the clutter $\{\{a\}, \{b\}\}$. Note that condition (C) holds and condition (C*) fails. Now $\tilde{\mathscr{C}}$ is a clutter consisting of one singleton set (which is in fact the set $\{I\}$) and so we have

$$P(\mathscr{C}, p) = 1 - (1-p)^2 > p = P(\tilde{\mathscr{C}}, p).$$

(b) Suppose now that $\mathscr{C}$ is the clutter $\{\{a, b\}\}$. Note that condition (C) fails but condition (C*) holds. Now $\tilde{\mathscr{C}}$ is in fact the same clutter as before, and so we have

$$P(\mathscr{C}, p) = p^2 < p = P(\tilde{\mathscr{C}}, p).$$

Theorem 2.1 (The clutter percolation theorem). *Let I be a finite non-empty set, let $\sim$ be an equivalence relation on I, let $\mathscr{C}$ be a clutter on I, and let $0 < p < 1$. Let*

$$\Delta = P(\mathscr{C}, p) - P(\tilde{\mathscr{C}}, p).$$

If condition (C) *holds, then $\Delta \geq 0$; if condition* (C*) *holds, then $\Delta \leq 0$; and so if both conditions* (C) *and* (C*) *hold, then $\Delta = 0$. Further, if exactly one of conditions* (C) *and* (C*) *hold, then $\Delta \neq 0$.*

The result that $\Delta \geq 0$ when condition (C) holds is related to a result used in [7] and [8].

3. Paths in random graphs and digraphs

Let us now look at some applications of the general theorem of the last section to the study of paths and connectedness in random graphs and digraphs. One result obtained concerns the existence of Hamiltonian cycles in large random digraphs. Other results may be of use for example in reliability theory and in the study of the vulnerability of communication and transportation networks (see for example Frank and Frisch [5]).

Firstly, what do we mean by random graphs and digraphs? We generalise the idea of the random graph $G_{n,p}$ and the random digraph $D_{n,p}$ introduced in Section 1. Let G be a (finite, undirected) graph and let $0 < p < 1$. We use G_p to denote the random subgraph of G which remains when we delete the edges of G independently with probability 1-p. Now recall that $D(G)$ denotes the digraph associated with G, which is obtained by replacing each edge of G by a pair of oppositely directed edges. We use $D(G)_p$ to denote the random subdigraph of $D(G)$ which remains when we delete the edges of $D(G)$ independently with probability 1-p.

In each of the applications below the set I will be the set of edges of $D(G)$ and $\sim$ will be the equivalence relation on I which makes equivalent the two edges (u, v) and (v, u) of $D(G)$ that arise from the edge $\{u, v\}$ of G. It will be convenient to identify the equivalence class $\{(u, v), (v, u)\}$ with the edge $\{u, v\}$ of G.

Let us note one preliminary result. By a *path* in a digraph we always mean a simple (without repeated vertices) directed path. If A is a path we denote its initial vertex by In A and its terminal vertex by Ter A. If v is a vertex on A we let $A \mid v$ denote the path up to v and $v \mid A$ denote the path from v on. The following simple observation will be very useful.

Lemma. *Let A and B be paths in $D(G)$, with the edge (u, v) in A and the edge (v, u) in B. Then from $A \mid u$ and $u \mid B$ we may form a path C with* In C = In A

and Ter C = Ter B; *and similarly from* $B \mid v$ *and* $v \mid A$ *we may form a path* D *with* In D = In B *and* Ter D = Ter A.

Given vertices s and t, we write "$s \rightsquigarrow t$" to mean that there is a path from s to t (in the appropriate graph or digraph).

Theorem 3.1 (see[6, 13]). $P\{s \rightsquigarrow t \text{ in } G_p\} = P\{s \rightsquigarrow t \text{ in } D(G)_p\}$.

Proof. Let $\mathscr{C}$ be the clutter of edge sets of simple s, t paths in $D(G)$. By the lemma $(\mathscr{C}, \sim)$ satisfies both conditions (C) and (C*). Further as noted earlier $\tilde{\mathscr{C}}$ is the clutter of edge sets of simple s, t paths in G. Hence by the clutter percolation theorem

$$P\{s \rightsquigarrow t \text{ in } G_p\} = P(\tilde{\mathscr{C}}, p) = P(\mathscr{C}, p) = P\{s \rightsquigarrow t \text{ in } D(G)_p\}.$$

Theorem 3.2. *$P\{G_p$ is Hamiltonian$\} \leq P\{D(G)_p$ is Hamiltonian$\}$, and further the inequality is strict if G is Hamiltonian.*

Proof. We may assume that there are at least three vertices. Let $\mathscr{C}$ be the clutter of edge sets of Hamiltonian cycles in $D(G)$. Clearly $(\mathscr{C}, \sim)$ satisfies condition (C), and fails to satisfy condition (C*) if G is Hamiltonian. Further, as noted earlier, $\tilde{\mathscr{C}}$ is the clutter of edge sets of Hamiltonian cycles in G. Hence by the clutter percolation theorem

$$\begin{aligned} P(G_p \text{ is Hamiltonian}) &= P(\tilde{\mathscr{C}}, p) \leq P(\mathscr{C}, p) \\ &= P(D(G)_p \text{ is Hamiltonian}), \end{aligned}$$

and the inequality is strict if G is Hamiltonian.

Recall that a digraph is *strongly connected* (or *di-connected*) if there is a path from each vertex to each other vertex.

Theorem 3.3. *$P(G_p$ is connected$) \geq P(D(G)_p$ is strongly connected$)$, and the inequality is strict if G is connected.*

Proof. Let $\mathscr{C}$ be the clutter of minimal edge sets of strongly connected subdigraphs of $D(G)$ (with the same vertex set). By the lemma $(\mathscr{C}, \sim)$ satisfies condition (C*), and it is easy to see that condition (C) fails if G is connected. Further $\tilde{\mathscr{C}}$ is the clutter of edge sets of spanning trees of G. Hence by the clutter percolation theorem

$$\begin{aligned} P(G_p \text{ is connected}) &= P(\tilde{\mathscr{C}}, p) \geq P(\mathscr{C}, p) \\ &= P(D(G)_p \text{ is strongly connected}), \end{aligned}$$

and the inequality is strict if G is connected.

For each positive integer n and $0 < p < 1$ write $D_{n,p}$ for $D(G)_p$ when G is the complete graph on n vertices (as in Section 1). Let α be a (small) constant and let $p = p(n) = (1+\alpha)(\log n)/n$, where the logarithm is natural.

Theorem 3.4. *As* $n \to \infty$,

$$P(D_{n,p} \text{ is Hamiltonian}) \to \begin{cases} 1 & \text{if } \alpha > 0, \\ 0 & \text{if } \alpha < 0. \end{cases}$$

This theorem improves on a result in Angluin and Valiant [1] and on other previous work, and answers a question Bondy [2].

Proof. Let $G_{n,p}$ denote the random graph on n (labelled) vertices in which the edges occur independently with probability p. By Theorem 3.2

$$P(D_{n,p} \text{ is Hamiltonian}) \geq P(G_{n,p} \text{ is Hamiltonian});$$

and by a result of Komlós and Szemerédi [10], if $\alpha > 0$ this last quantity tends to 1 as $n \to \infty$. Conversely, by Theorem 3.3

$$\begin{aligned} P(D_{n,p} \text{ is Hamiltonian}) &\leq P(D_{n,p} \text{ is strongly connected}) \\ &\leq P(G_{n,p} \text{ is connected}); \end{aligned}$$

and by a result of Erdös and Renyi (see [10]) if $\alpha < 0$, this last quantity tends to 0 as $n \to \infty$. (Alternatively the second part follows simply from looking at vertex degrees.)

Let us note two further examples of results similar to those above which may be deduced from the clutter percolation theorem in much the same way. We say (here) that a graph or digraph is *k-connected* if for every pair of vertices u, v there are k edge-disjoint paths from u to v.

Theorem 3.5. *For any positive integer k,*

$$P(G_p \text{ is } k\text{-connected}) \geq P(D(G)_p \text{ is } k\text{-connected}),$$

and the inequality is strict if G is k-connected.

The *distance* from a vertex u to a vertex v is the least number of edges in a path from u to v (and is ∞ if there is no such path). The *diameter* of a graph or digraph is the greatest distance between two vertices.

Theorem 3.6. *For any positive integer d,*

$$P(G_p \text{ has diameter} \leq d) \geq P(D(G)_p \text{ has diameter} \leq d)$$

and the inequality is strict if G has diameter d.

Note that when $d+1$ is at least the number of vertices of G or D we obtain the main part of Theorem 3.3.

4. Two results on percolation

In this section we deduce easily from the clutter percolation theorem two important results in "classical" percolation theory which had previously been given quite different ad hoc proofs.

4.1. *Atom and bond percolation*

Let G be a (finite) partially directed graph, with a (source) vertex s and a set T of (sink) vertices. Suppose that the vertices (or atoms) other than s are open with probability p_a (and otherwise are closed or blocked) and the edges (or bonds) are open with probability p_b, and that these events occur independently. We obtain a random graph G_{p_a,p_b} say, and we are interested in the probability that there is an unblocked path from s to T in this random graph. Let us denote this probability by $P(s \rightsquigarrow T \text{ in } G_{p_a,p_b})$. "Atom percolation" is case $p_b = 1$ and "bond percolation" the case $p_a = 1$.

Theorem 4.1 (Hammersley [7]). *If* $0 < p < 1$, *then*

$$P(s \rightsquigarrow T \textit{ in } G_{p,1}) \leq P(s \rightsquigarrow T \textit{ in } G_{1,p}).$$

Proof. Note first that we may assume that G is completely directed. For by a variant of Theorem 3.1

$$P(s \rightsquigarrow T \textit{ in } G_{p_a,p_b}) = P(s \rightsquigarrow T \textit{ in } D(G)_{p_a,p_b})$$

where $D(G)$ is the associated digraph.

Let I be the set of edges of G and let $\sim$ be the equivalence relation on I which makes equivalent edges with the same "head" (so that $(u, w) \sim (v, w)$). Let us identify with w the equivalence class of all edges with head w. Let $\mathscr{C}$ be the clutter of minimal edge sets of paths in G from s to T. Then $\tilde{\mathscr{C}}$ is the clutter of minimal vertex sets of paths in G from s to T, each less its initial vertex s. Also $(\mathscr{C}, \sim)$ satisfies condition (C). Hence by the clutter percolation theorem

$$\begin{aligned} P(s \rightsquigarrow T \textit{ in } G_{p,1}) &= P(\tilde{\mathscr{C}}, p) \leq P(\mathscr{C}, p) \\ &= P(s \rightsquigarrow T \textit{ in } G_{1,p}). \end{aligned}$$

We may deduce easily from Theorem 4.1 an apparently more general result.

Corollary 4.2. *If* $0 < p_1 < p_2 \leq 1$, *then*

$$P(s \rightsquigarrow T \textit{ in } G_{p_1,p_2}) \leq P(s \rightsquigarrow T \textit{ in } G_{p_2,p_1}).$$

Proof. By thinking of G_{p_1,p_2} and G_{p_2,p_1} as being formed in two stages we see that (putting $p = p_1/p_2$),

$$P(s \rightsquigarrow T \text{ in } G_{p_1,p_2}) =$$

$$= \sum_H P(G_{p_2,p_2} = H) P(s \rightsquigarrow T \text{ in } H_{p,1})$$

$$\leq \sum_H P(G_{p_2,p_2} = H) P(s \rightsquigarrow T \text{ in } H_{1,p})$$

$$= P(s \rightsquigarrow T \text{ in } G_{p_2,p_1}).$$

Let us say that G is *treelike* with respect to s and T if any two minimal paths from s to T with a common vertex are identical up to that vertex (see Fig. 1). If G is treelike it is easy to see that we must have equality in the results above. If G is not treelike, then the relevant condition (C*) above fails. Hence by the clutter percolation theorem we have strict inequality in Theorem 4.1 (as shown in [7]) and so also in the corollary.

Fig. 1.

4.2. *Percolation and Bethe trees*

As in the last application let G be a (finite) partially directed graph with a (source) vertex s and a set T of (sink) vertices. The corresponding *Bethe tree* $\hat{G}$ is useful in calculations concerning percolation probabilities (see [8]). It has a vertex v_σ for each simple path σ in G starting at s, and if τ is a continuation of σ by one further edge then v_σ is joined to v_τ. Thus $\hat{G}$ is a tree rooted at $\hat{s}$ say, where $\hat{s}$ is the vertex corresponding to the trivial path in G at s. Let $\hat{T}$ be the set of vertices of $\hat{G}$ corresponding to paths in G from s to T. Let $0<p<1$ and suppose that the edges of G and $\hat{G}$ are open independently with probability p, yielding random graphs G_p and $\hat{G}_p$.

Theorem 4.3 (Hammersley and Walters [8]).

$$P\{s \rightsquigarrow T \text{ in } G_p\} \leq P\{\hat{s} \rightsquigarrow \hat{T} \text{ in } \hat{G}_p\}.$$

Proof. Let I be the set of edges of $\hat{G}$ and let $\sim$ be the equivalence relation on I which makes equivalent edges arising from a common edge of G. Let us identify an equivalence class with the corresponding edge of G. Let $\mathscr{C}$ be the clutter of minimal edge sets of paths in $\hat{G}$ from $\hat{s}$ to $\hat{T}$. Then condition (C) holds, and $\tilde{\mathscr{C}}$ is the clutter of minimal edge sets of paths in G from s to T. Hence by the clutter percolation theorem

$$P\{\hat{s} \rightsquigarrow \hat{T} \text{ in } \hat{G}_p\} = P(\mathscr{C}, p)$$
$$\geq P(\tilde{\mathscr{C}}, p) = P\{s \rightsquigarrow T \text{ in } G_p\}.$$

Let us say that G is *weakly treelike* (with respect to s and T) if any two minimal paths from s to T with a common edge are identical up to that edge (see Fig. 2). If G is weakly treelike, then it is easy to see that we must have equality in Theorem 4.3 above. If G is not weakly treelike, then in the proof above the relevant condition (C*) fails, and so we have strict inequality above (as shown in [8]).

In a recent paper, Oxley and Welsh [12], there is given a straightforward inductive proof of Theorem 4.1, and an inductive proof of Theorem 4.3 which may be thought of as being based on a lemma in Harris [9]. This lemma of Harris (which is a special case of the FKG inequality [4]) also follows easily from the clutter percolation theorem.

Fig. 2.

Acknowledgment

I would like to thank Laurence Wolsey for helpful comments on the preparation of this paper.

References

[1] D. Angluin and L.G. Valiant, "Fast probabilistic algorithms for Hamiltonian circuits and matchings", Internal report CSR-17-77, University of Edinburgh (1977).

[2] J.A. Bondy, "Hamiltonian cycles in graphs and digraphs", Research Report CORR 78-16, University of Waterloo, Canada (1978).

[3] J.A. Bondy and U.S.R. Murty, *Graph theory with applications* (Macmillan Press, London, 1977).

[4] C.M. Fortuin, P.W. Kasteleyn and J. Ginibre, "Correlation inequalities on some partially ordered sets", *Communications of Mathematical Physics* 22 (1971) 89-103.

[5] H. Frank and I.T. Frisch, *Communication, transmission and transportation networks* (Addison-Wesley, New York, 1971).

[6] H.L. Frisch and J.M. Hammersley, "Percolation processes and related topics", *Journal of the Society for Industrial and Applied Mathematics* 11 (1963) 894–918.

[7] J.M. Hammersley, "Comparison of atom and bond percolation processes", *Journal of Mathematical Physics* 2 (1961) 728–733.

[8] J.M. Hammersley and R.S. Walters, "Percolation and fractional branching processes", *Journal of the Society for Industrial and Applied Mathematics* 11 (1963) 831–839.

[9] T.E. Harris, "A lower bound for the critical probability in a certain percolation process", *Proceedings of the Cambridge Philosophical Society* 56 (1960) 13–20.

[10] J. Komlós and E. Szemerédi, "Limit distribution for the existence of Hamilton cycles in a random graph" (to appear).

[11] C.H. McDiarmid, "General percolation and random graphs", *Advances in Applied Probability*, to appear.

[12] J.G. Oxley and D.J.A. Welsh, "On some percolation results of J.M. Hammersley", *Journal of Applied Probability*, to appear.

[13] A. Satyanarayana and A. Prabhakar, "New topological formula and rapid algorithm for reliability analysis of complex networks", *IEEE Transactions in Reliability* R27 (1978) 82–100.

Mathematical Programming Study 13 (1980) 26–34.
North-Holland Publishing Company

THE USE OF RECURRENCE RELATIONS IN COMPUTING

L.B. WILSON

University of Stirling, Stirling, Scotland

Received 1 February 1980

The theme of this paper is that recurrence relations play an important part in computing science. Several examples are given in enumeration, systematic ordering, and the analysis of algorithms to illustrate this contention.

Key words: Analysis of Algorithms, Binary Sequence Search Trees (BSST), Binary Sequences, Derangements, Difference Equations, Recurrence Relations, Restricted Permutations, Systematic Orderings.

1. Introduction

In this paper selected examples have been taken from different areas of computing to highlight the use of recurrence relations. The terms "recurrence relation" and "difference equation" are synonymous the former is more commonly used in computing and the latter in mathematics. The most obvious use of recurrence relations is in the enumeration of the members of sets and in Section 2 we examine some straightforward combinatorial problems of this type. Less well-known is their use in the systematic generation of ordered sequences and an example of this type is given in Section 3.

The importance of the analysis of algorithms in Computing is now generally accepted and the books by Aho, et al. [1] and Knuth [4, 5, 6] have been pioneers in this field. However, it is perhaps not fully realised how necessary the ability to handle and solve recurrence relations is to the successful analysis of many algorithms and in Section 4 a complete example is given to show this.

2. The enumeration of sets

Recurrence relations have often been found to be convenient methods for enumerating the members of a set and there are many examples of their use in this way. We will look at two such examples in this section.

2.1. Binary sequences

The following simple problem is typical of those in the enumeration of sets:

What is the number of binary sequences of length n which do not contain two consecutive one's?

Let the number of such sequences be W_n, and let U_n and V_n be the number of such sequences whose last digit is a one or a zero respectively. Consider extending sequences of this type from length $n-1$ to n we have two possible situations

(1) If a correct $n-1$ sequence ends with a one we can append a zero but not another one.

(ii) If a correct $n-1$ sequence ends with a zero we can append either a zero or a one.

Thus we obtain the two recurrence relations

$$V_n = V_{n-1} + U_{n-1}, \tag{1}$$

$$U_n = V_{n-1} \tag{2}$$

which lead directly to the equations

$$V_n = V_{n-1} + V_{n-2},$$

$$U_n = U_{n-1} + U_{n-2}$$

which when added give the recurrence relation for our problem

$$W_n = W_{n-1} + W_{n-2}. \tag{3}$$

This equation is the same as that obtained for Fibonacci numbers, and being linear with constant coefficients is easily solved using the initial conditions $W_1 = 2$ and $W_2 = 3$.

The crucial idea in this example is to divide the original set into two mutually exclusive subsets; a binary sequence must end either with a one or a zero but not both. We therefore count these occurrences and add them together. The difficulty in such problems depends on the ease with which we can find ways of dividing the set into mutually exclusive and exhaustive subsets. Mutual exclusion means we do not count the items twice and exhaustion means we do not fail to count some item. Further examples of this technique are to be found in a recent book by Page and Wilson [9]. An interesting method of deriving such recurrence relations using finite state grammars to define a suitable language is given in a paper by Cohen and Katcoff [2], however, their method seems more of a novelty than a practical alternative.

2.2. *Derangements*

A derangement of the marks $a_1, a_2, \ldots, a_n$ is a permutation of them such that no mark remains in its original position, i.e. a_1 is not in the first position, a_2 in the second position etc. Therefore 365124 is a derangement of the first six integers but 536412 is not. Enumeration of the set of derangements using recurrence relations can be done as follows: Let D_n be the number of derangements of the n

integers $(1, 2, \ldots, n)$. Consider the first position to be occupied by the integer k $(k \neq 1)$. Now the displaced integer 1 can either be in the kth position or not. If it is in the kth position, then we have the derangements of the $n-2$ integers $2, 3, \ldots, k-1, k+1, \ldots, n$, i.e. D_{n-2}. If it is not in the kth position, then the kth position can be considered as the forbidden position for the integer 1 and we have the derangements of all the n integers except k i.e. D_{n-1}. Since k can assume any of the $n-1$ values $2, 3, \ldots, n$ we have the recurrence relation

$$D_n = (n-1)(D_{n-1} + D_{n-2}). \tag{4}$$

This equation is not quite so easy to solve as (3) although it is still linear it no longer has constant coefficients. We can, however, rearrange (4) as follows

$$D_n - nD_{n-1} = -D_{n-1} + (n-1)D_{n-2}. \tag{5}$$

If we let $F_n = D_n - nD_{n-1}$, then (5) becomes $F_n = -F_{n-1}$, which can be simply solved giving $F_n = (-1)^n$ using the initial condition $F_2 = 1$. Thus we have reduced the problem to solving the first-order difference equation

$$D_n - nD_{n-1} = (-1)^n. \tag{6}$$

Using standard methods (see for example Page and Wilson [9]) the solution of (6) is

$$D_n = Cn! + n! \sum_{i=1}^{n-1} \frac{(-1)^{i+1}}{(i+1)!}$$

where C is an arbitrary constant which can be found from the initial condition $D_1 = 0$, giving $C = 0$.

3. Systematic orderings

In the previous section we saw how recurrence relations were used to enumerate sets but their use in the following combinatorial processes is less well-known

(i) List the members of a set in some systematic order.
(ii) Given a member find its position in the order.
(iii) Given a position find the member which occupies it.
(iv) Select at random a member of the set.

Page [8] gave some examples of the use of recurrence relations in such areas of combinatorics and we will examine one such example in this section.

3.1. *Restricted permutations with repetition*

Consider the r-permutations of the n-objects $(1, 2, \ldots, n)$ with unlimited repetition but with the restriction that no three adjacent objects are the same.

Let $\mathcal{P}_r^n$ be the set of such permutations and P_r^n the number of members of this set. In the usual way we divide these permutations into two mutually exclusive subsets depending on whether the last two symbols are the same or different. If they are the same then the r-permutation can be obtained from any (r-2)-permutation by attaching two like symbols to it providing these symbols are distinct from the last one of the $(r-2)$-permutation, i.e. $(n-1)P_{r-2}^n$. Turning now to the r-permutations whose last two symbols are different they can be obtained by attaching a different symbol at the end of the $(r-1)$-permutation, i.e. $(n-1)P_{r-1}^n$. Thus the basic recurrence relation is

$$P_r^n = (n-1)(P_{r-2}^n + P_{r-1}^n). \tag{7}$$

The initial conditions are $P_1^n = n$, $P_2^n = n^2$, and we can now solve the recurrence relation (7) to find P_r^n since it is linear with constant coefficients.

The systematic ordering is derived naturally from the recurrence relation (7) and so the ordering for r-permutations involves both $(r-1)$ and $(r-2)$-permutations. For example to obtain the permutations $\mathcal{P}_3^3$ we need the appropriate permutations for $r=1$ and $r=2$ and these are (we consider $n=3$).

$\mathcal{P}_1^3$	1		2		3				
$\mathcal{P}_2^3$	11	12	13	21	22	23	31	32	33

The permutations $\mathcal{P}_3^3$ are written down systematically by attaching all possible identical pairs of objects in turn to the $\mathcal{P}_1^3$ members followed by adding all possible single objects in turn to the $\mathcal{P}_2^3$ members. This gives the following permutations (reading row by row)

$\mathcal{P}_3^3$	122	133	211	233	311	322
	112	113	121	123	131	132
	212	213	221	223	231	232
	312	313	321	323	331	332

Similarly we can obtain $\mathcal{P}_r^3$ for $r > 3$.

The identification of the place in the ordering occupied by a given r-permutation of this restricted type is found by successively determining whether it is derived from an $(r-1)$ or an $(r-2)$-permutation and so on until permutations with one or two objects are obtained. We can then work back and calculate the position of the given r-permutation. Let V_r be the position of the r-permutation and its last three digits be x, y and z, then the following formulae hold

if $y = z$, then we obtained $\mathcal{P}_r^n$ from $\mathcal{P}_{r-2}^n$ and

$$V_r = (V_{r-2} - 1)(n-1) + \textbf{if } z > x \textbf{ then } z - 1 \textbf{ else } z; \tag{8}$$

if $y \neq z$, then we obtained $\mathcal{P}_r^n$ from $\mathcal{P}_{r-1}^n$ and

$$V_r = P_{r-2}^n \times (n-1) + (V_{r-1} - 1)(n+1) + \textbf{if } z > y \textbf{ then } z - 1 \textbf{ else } z. \tag{9}$$

Let us take an example to illustrate the process. For $n = 3$ find the position of the 5-permutation 31223.

(1) The last two digits are unequal so we obtained it from 3122 with 3 added.

(2) 3122 has equal end digits so it came from 31 with 22 added.

(3) 31 is the seventh permutation in $\mathcal{P}_2^3$.

Working back: $V_2 = 7$,

From (8) $V_4 = (V_2 - 1)(n-1) + 1 = 13$,

From (9) $V_5 = P_3^3 \times (n-1) + (V_4 - 1)(n-1) + 2 = 74$.

The reverse process of finding the permutation at a given place in the order also uses (8) and (9). We repeatedly divide by $(n-1)$, subtracting P_{r-2}^n if the quotient exceeds it and noting the remainders. In this way we can build up the permutation from either a member of $\mathcal{P}_1^n$ or $\mathcal{P}_2^n$.

Further details of these techniques can be found in Page [8] and Page and Wilson [9].

4. Analysis of algorithms

Although the two previous sections have shown how to use recurrence relations in computing their most important application is in analysing algorithms. When we examine the classic books on the analysis of algorithms by Aho et al. [1] and Knuth [4, 5, 6] we observe wide and varied use of recurrence relations. Let us examine a typical example.

4.1. *The analysis of binary sequence search trees*

The method of searching using a binary sequence search tree (BSST) is well-known being first suggested by Hibbard [3]. Given a list of keys the binary tree is constructed by making the first item the root and a subsequent item is placed on the left if its key is less than the current node and on the right if it is greater. We place the item at the first unoccupied node. For example the keys 2413 would give the BSST given in Fig. 1.

In this figure we have drawn a square box where there is a vacant space below a node. These square boxes represent all the places where a new node can be placed and we can prove by induction that a BSST with n nodes has $(n+1)$ such square box positions.

The number of comparisons to find a node when searching the BSST is equivalent to the number required to insert that node originally. A typical tree building routine would be

```
procedure build tree (reference (node) value T; integer value x);
if T = null then T := node(x, null, null)
        else if x > data(t) then build tree (rlink(T), x)
                             else build tree (llink(T), x);
```

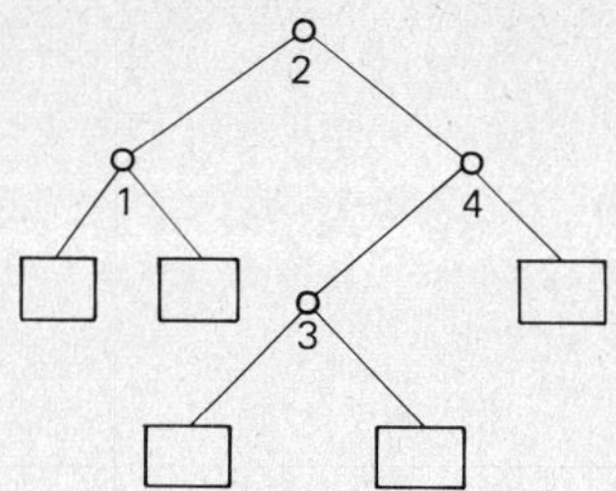

Fig. 1. The BSST for 2413.

When we analyse searching using a BSST it is important to consider all the possible $n!$ permutations of $1, 2, \ldots, n$ and not all the possible BSST's. This is because several permutations can produce the same BSST (e.g. 31542, 35142, 35124, 31524) whilst the permutation 12345 is the only one to give its BSST. We can carry out a simple analysis using a recurrence relation in U_n, the number of comparisons required to build a BSST with n nodes averaged over all permutations of $1, 2, \ldots, n$ (see for example Wilson [10]). However, such an analysis only finds the average number of comparisons and if we are interested in the variance we need a more sophisticated approach using a two-variable recurrence relation.

Let A_{nk} be the number of permutations of $1, 2, \ldots, n$ whose last element requires k comparisons[1] to insert. In order to derive a recurrence relation for A_{nk} each permutation of the $(n-1)$ objects $a_1a_2 \ldots a_{n-1}$ needs to be extended into n permutations of n objects. There are several methods of doing this, the one we use here is to insert b, where b takes all values $1 \leq b \leq n$, in the second last place of the original permutation. So we have $a'_1a'_2 \ldots a'_{n-2}ba'_{n-1}$ and $a'_i = a_i$ if $a_i < b$ and $a'_i = a_i + 1$ if $a_i \geq b$.

For example consider the permutation 2413 from which Fig. 1 was obtained. It has a k value of 3 and when it is extended by the method described above we obtain the five permutations 35214, 35124, 25134, 25143, 24153 for $b = 1, 2, 3, 4, 5$ respectively. The BSST's for these five permutations are given in Fig. 2, and we can see that these five trees are equivalent to replacing in turn each one of the square boxes in the original tree of Fig. 1.

The k values for the five permutations are $3, 3, 4, 4, 3$ and in general the k value of any n-permulation obtained by this construction will remain the same value as the $n-1$ permutation it came from unless $b = a_{n-1}$ or $b = a_{n-1} + 1$ when k will be increased by one. In this example since $a_{n-1} = 3$ the values of b which give an increased k are $b = 3, 4$.

[1] Strangely enough such a simple idea as the number of comparisons is not universally agreed. We consider that the root (at level 1) requires one comparison to find that the location is empty, and subsequent items at level k require k comparisons. Other authors will have one less comparison in all cases and hence a slightly different final answer.

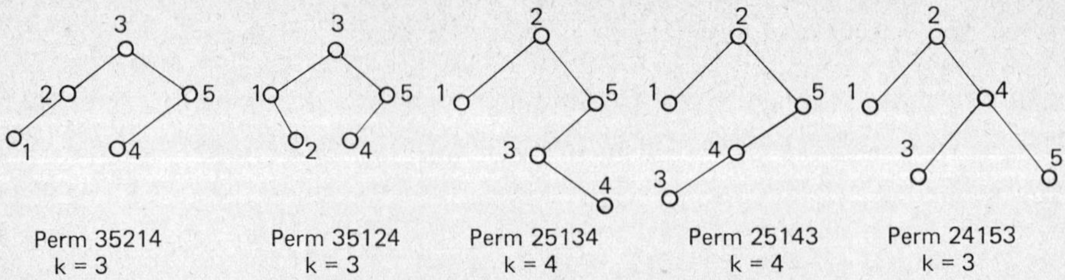

Fig. 2. The five BSST's obtained by extending perm 2413.

Therefore in general

$$A_{nk} = (n-2)A_{n-1,k} + 2A_{n-1,k-1}. \tag{10}$$

The first term on the right-hand side are those $n-2$ cases where k does not change and the other term the two cases in which k has been increased. There are other ways of deriving the recurrence relation (10) based on the fact that when a new node is inserted one square box is replaced by two square boxes at one level further down the BSST.

The boundary conditions for the recurrence relation (10) are

$$A_{11} = 1,\ A_{1k} = 0\ (k \neq 1),\ A_{nk} = 0\ (k > n).$$

One method of solving the recurrence relation is to use a generating function

$$G_n(x) = \sum_k A_{nk}x^k$$

substituting for A_{nk} from (10) gives

$$\begin{aligned} G_n(x) &= \sum_k [(n-2)A_{n-1,k} + 2A_{n-1,k-1}]x^k \\ &= (n-2)\sum_k A_{n-1,k}x^k + 2x\sum_k A_{n-1,k-1}x^{k-1} \\ &= (n-2)G_{n-1}(x) + 2xG_{n-1}(x) = (n-2+2x)G_{n-1}(x). \end{aligned}$$

Therefore

$$G_n(x) = (n-2+2x)(n-3+2x)\ldots(0+2x)G_1(x)$$

and

$$G_1(x) = \sum_k A_{1k}x^k = A_{11}x = x.$$

Consider now the probabilities P_{nk}, where P_{nk} is the probability that a random permutation of n elements requires k comparisons to insert the last item.

$$P_{nk} = A_{nk}/n!,$$

$$P_n(x) = \sum_k P_{nk}x^k = \frac{G_n(x)}{n!} = \left(\frac{n-2+2x}{n}\right)\left(\frac{n-3+2x}{n-1}\right)\cdots\left(\frac{0+2x}{2}\right)\frac{x}{1}.$$

The average number of comparisons is:

$$P_n'(1) = \frac{2}{n} + \frac{2}{n-1} + \dots + \frac{2}{2} + \frac{1}{1} = 2H_n - 1. \tag{11}$$

The variance of the number of comparisons is:

$$\begin{aligned} &P_n''(1) + P_n'(1) - (P_n'(1))^2 = \\ &= \left(\frac{2}{n} - \frac{4}{n^2}\right) + \left(\frac{2}{n-1} - \frac{4}{(n-1)^2}\right) + \dots + \left(\frac{2}{3} - \frac{4}{3^2}\right) + \left(\frac{2}{2} - \frac{4}{2^2}\right) \\ &= 2H_n - 4H_n^{(2)} + 2 \end{aligned}$$

where

$$H_n = \sum_{i=1}^{n} \frac{1}{i} \quad \text{and} \quad H_n^{(2)} = \sum_{i=1}^{n} \frac{1}{i^2}. \tag{12}$$

The worst case analysis of this problem gives a very bad result (caused by examples such as 1 2 3 4 ... $n-1$ n) and so it is interesting to note that the variance given by (12) is fairly stable about the mean.

5. Discussion

Recurrence relations (or difference equations) have like many other aspects of combinatories dropped out of the normal mathematics syllabus, so much so that one of the few good books on the subject by Milne-Thomson [7] was first published over 40 years ago. Subjects seem to go through fashions and the current "in" subject in Combinatorics is Graph Theory. From the practical point of view recurrence relations occupy an important position. In many ways they are equivalent in discrete mathematics to the position of differential equations in continuous mathematics. However, differential equations are widely taught and there is a considerable amount of research in this area.

We have seen that recurrence relations are important to problem solving in Computing but they have wider implications in computing when considered in conjunction with recursion. Recursion is a fundamental concept in computing which has not yet received sufficient attention, partly because it is much more difficult to comprehend than iteration and requires considerable effort from both the student and teacher to understand it. Once it has been understood it can be used both as a programming technique and as a method of problem solving. Many problems are soluble by considering how the n case can be derived from $n-1$, $n-2$, ... cases, and this fairly naturally leads on to a recurrence relation.

It is hoped that the examples given in this paper, brief though they are, have shown the importance of recurrence relations and will help to revive an interest in them from both mathematicians and computing scientists. It is a topic, which together with elementary configurations such as permutations and trees, counting, ordering, and generating functions, should be taught to computing scientists emphasing more the methods and less the mathematical theorems.

References

[1] A.V. Aho, J.E. Hopcroft and J.D. Ullman, *The design and analysis of computer algorithms* (Addison-Wesley, Reading, MA, 1974).

[2] J. Cohen and J. Katcoff, "Automatic solution of a certain class of combinatorial problems", *Information Processing Letters* 6 (1977) 101–104.

[3] T.N. Hibbard, "Some combinatorial properties of certain trees with applications to searching and sorting", *Journal of the Association of Computing Machinery* 9 (1962) 13–29.

[4] D.E. Knuth, *The art of computer programming. Volume* 1: *fundamental algorithms* (Addison-Wesley, Reading, MA, 1968).

[5] D.E. Knuth, *The art of computer programming. Volume* 2: *seminumerical algorithms* (Addison-Wesley, Reading, MA, 1969).

[6] D.E. Knuth, *The art of computer programming. Volume* 3: *sorting and searching* (Addison-Wesley, Reading, MA, 1973).

[7] L.M. Milne-Thomson, *The calculus of finite differences* (originally published 1933 but republished in 1951 by Macmillan, London).

[8] E.S. Page, "Systematic generation of ordered sequences using recurrence relations", *Computing Journal* 14 (1971) 150–153.

[9] E.S. Page and L.B. Wilson, *An introduction to computational combinatorics* ((Cambridge University Press, Cambridge, 1979).

[10] L.B. Wilson, "Sequence search trees: their analysis using recurrence relations", *BIT* 16 (1976) 332–337.

Mathematical Programming Study 13 (1980) 35–52.
North-Holland Publishing Company

A BRANCH AND BOUND ALGORITHM FOR THE KOOPMANS–BECKMANN QUADRATIC ASSIGNMENT PROBLEM

C.S. EDWARDS

Department of Engineering Production, University of Birmingham, Birmingham, Great Britain

Received 1 February 1980

In this paper a binary branch and bound algorithm for the exact solution of the Koopmans–Beckmann quadratic assignment problem is described which exploits both the transformation and the greedily obtained approximate solution described in a previous paper by the author. This branch and bound algorithm has the property that at each bound an associated solution is obtained simultaneously, thereby rendering any premature termination of the algorithm less wasteful.

Key words: Branch and Bound, Greedy Approximation, Quadratic Assignment.

1. Introduction

In CP77 [1], I described in detail a transformation of the objective function of the Koopmans–Beckmann Quadratic Assignment Problem (copies of the paper in CP77 can also be obtained from the author at the above address). As in the earlier paper, henceforward in this present paper this problem will be called "the K.–B.P.". The transformation of the objective function referred to above will be called "the canonical form of the K.–B.P.".

In this paper we show how the canonical form of the K.–B.P. can be exploited within a branch and bound algorithm to find an exact solution of the problem; alternatively, if time does not permit completion of the algorithm, then this can be truncated at any time after a very early stage with already determined upper and lower bounds to whichever of the two extremal values of the objective function is of interest to us; always in this paper it will be supposed that our interest is in the minimum value of the objective function.

In our algorithm the bounding procedure essentially is that used 16 or more years ago independently by both Lawler [4] and Gilmore [2]; however, here the procedure is applied more efficiently in that it is applied to the "minimal" quadratic residual terms in the canonical form of the K.–B.P.; also, a binary branching procedure is preferred to the n-tuple branching used by Lawler and by Gilmore.

In our preferred minimising variant of the K.–B.P. we shall use, amongst other upper bounding procedures, the greedy procedure described in [1]. Indeed this present paper should be regarded as a sequel to the latter paper from which both notation and intermediate results will be quoted extensively in order not to prolong unduly the length of this paper. For the same reason the applicability of the K.–B.P. and the difficulty of finding extremal values of the corresponding objective function will not be discussed here; the applicability and the difficulty are both well-known and are discussed in detail, for example, in the references listed in [1].

2. Notation

For convenience, we now give a summary of the matrix notation used in this paper; this notation is consistent with that used in [1].

Unless otherwise stated each boldface capital letter denotes a square matrix of order n (here called an n-matrix), where n is an arbitrarily chosen integer ≥ 1; if the capital letter used is Greek, then the boldface capital letter denotes a diagonal matrix of order n, i.e. a square matrix of which at most the principal diagonal elements are non-zero.

Example. $\boldsymbol{X}$ denotes an n-matrix. We write $\boldsymbol{X} = [x_{ij}]$ where we wish to convey the information that x_{ij} denotes the element in row i and column j of $\boldsymbol{X}$ ($1 \leq i \leq n$ and $1 \leq j \leq n$).

If $\boldsymbol{X} = [x_{ij}]$ is any n-matrix, then $\operatorname{tr} \boldsymbol{X} \overset{\text{def}}{=} \sum_{i=1}^{n} x_{ii}$; if $\boldsymbol{X}$ has an inverse, then $\boldsymbol{X}^{-1}$ denotes this inverse.

Unless otherwise stated, each boldface lower case letter denotes an $(n \times 1)$ matrix, i.e. a matrix with n rows and 1 column.

Example. $\boldsymbol{y} = (y_1, y_2, \dots, y_n)$ denotes an $(n \times 1)$ matrix of which the element in row i is y_i ($1 \leq i \leq n$).

If $\boldsymbol{X} = [x_{ij}]$ is any n-matrix, then $\boldsymbol{X}'$ denotes the n-matrix of which the element in row i and column j is x_{ji}. If $\boldsymbol{y} = (y_1, y_2, \dots, y_n)$, then $\boldsymbol{y}'$ notes the $(1 \times n)$ matrix (i.e. a matrix with 1 row and n columns), of which the element in column j is y_j ($1 \leq j \leq n$), and we write $\boldsymbol{y}' = [y_1, y_2, \dots, y_n]$. $\boldsymbol{X}'$ and $\boldsymbol{y}'$ will be called "the transpose of $\boldsymbol{X}$" and the "the transpose of $\boldsymbol{y}$", respectively.

$\boldsymbol{P} = [p_{ij}]$ denotes any one of the $n!$ permutation matrices of order n. (A permutation matrix has precisely one element in each row and in each column which is 1, whilst every other element is 0; a permutation matrix is necessarily square and, if $\boldsymbol{P}$ is any such matrix, then $\boldsymbol{P}^{-1} = \boldsymbol{P}'$ and so $\boldsymbol{P}$ is an orthogonal matrix.)

$\boldsymbol{I}_n$ denotes the $(n \times n)$ identity matrix, $\boldsymbol{J}_n$ denotes the n-matrix of which each element is 1, and $\boldsymbol{O}_n$ denotes the n-matrix of which each element is zero; $\boldsymbol{h}_n$ and $\boldsymbol{o}_n$ denote the $(n \times 1)$ matrices of which each element is 1 and 0, respectively.

Where no ambiguity results, $\boldsymbol{I}_n$, $\boldsymbol{J}_n$, $\boldsymbol{O}_n$, $\boldsymbol{h}_n$ and $\boldsymbol{o}_n$ may be written $\boldsymbol{I}$, $\boldsymbol{J}$, $\boldsymbol{O}$, $\boldsymbol{h}$ and $\boldsymbol{o}$, respectively.

In [1] I defined two matrix transformations and gave references to earlier uses of these or similar transformations. We can give these definitions as follows:

If $\boldsymbol{X} = [x_{ij}]$ is any n-matrix, then

$$\boldsymbol{X}^* = [x_{ij}^*] \stackrel{\text{def}}{=} \left(\boldsymbol{I} - \frac{1}{n}\boldsymbol{J}\right)\boldsymbol{X}\left(\boldsymbol{I} - \frac{1}{n}\boldsymbol{J}\right).$$

If $\boldsymbol{W} = [w_{ij}]$ is any n-matrix, $n > 2$, such that each $w_{ii} = 0$ $(1 \le i \le n)$, then

$$\boldsymbol{W}^{**} = [w_{ij}^{**}] \stackrel{\text{def}}{=} \boldsymbol{W} - \boldsymbol{\Omega} - \frac{((n-1)\boldsymbol{W} + \boldsymbol{W}')\boldsymbol{J}}{n(n-2)} - \frac{\boldsymbol{J}((n-1)\boldsymbol{W} + \boldsymbol{W}')}{n(n-2)} + \frac{\boldsymbol{JWJ}}{(n-1)(n-2)},$$

where $\boldsymbol{\Omega}$ is the uniquely determined diagonal matrix such that each $w_{ii}^{**} = 0$ $(1 \le i \le n)$.

Note. $w_{11}, w_{22}, \ldots, w_{nn}$ and $w_{11}^{**}\ w_{22}^{**}, \ldots, w_{nn}^{**}$ are the principal diagonal elements (p.d.e.) of $\boldsymbol{W}$ and $\boldsymbol{W}^{**}$, respectively.

In [1] we find the following almost trivial result:

Lemma 1.

$$\boldsymbol{X}^* = \boldsymbol{X} - \frac{1}{n}\boldsymbol{XJ} - \frac{1}{n}\boldsymbol{JX} + \frac{1}{n^2}\boldsymbol{JXJ}$$

and $\boldsymbol{X}^*$ has each row sum and column sum zero.

Remark. The first part of this lemma is equivalent to the statement that

$$x_{ij}^* = x_{ij} - \frac{1}{n}\sum_{j=1}^{n} x_{ij} - \frac{1}{n}\sum_{i=1}^{n} x_{ij} + \frac{1}{n^2}\sum_{i=1}^{n}\sum_{j=1}^{n} x_{ij}$$

where, clearly, the 2nd, 3rd and 4th terms on the right-hand side are the mean element in row i, the mean element in column j and the overall mean element in $\boldsymbol{X}$, respectively.

Now, any n-matrix $\boldsymbol{X}$ can be expressed in a unique way as the sum of a symmetric n-matrix and a skew-symmetric n-matrix (a symmetric matrix is equal to its transpose and a skew-symmetric matrix is equal to minus its transpose). $\frac{1}{2}(\boldsymbol{X} + \boldsymbol{X}')$ and $\frac{1}{2}(\boldsymbol{X} - \boldsymbol{X}')$ are called the symmetric component of $\boldsymbol{X}$ and the skew-symmetric component of $\boldsymbol{X}$, respectively.

If $\boldsymbol{W}$ and $\boldsymbol{D}$ are n-matrices with each p.d.e. zero, then

$$\boldsymbol{A} \stackrel{\text{def}}{=} \tfrac{1}{2}(\mathbf{W} + \mathbf{W}'), \qquad \boldsymbol{B} \stackrel{\text{def}}{=} \tfrac{1}{2}(\mathbf{D} + \mathbf{D}'), \qquad \boldsymbol{F} \stackrel{\text{def}}{=} \tfrac{1}{2}(\boldsymbol{W} - \boldsymbol{W}'), \qquad \boldsymbol{G} \stackrel{\text{def}}{=} \tfrac{1}{2}(\boldsymbol{D} - \boldsymbol{D}').$$

We see that both A and F have each p.d.e. zero and the following result is obtained directly from the definition of W^{**}:

$$A^{**} = A - \Omega - \frac{1}{n-2} AJ - \frac{1}{n-2} JA + \frac{JAJ}{(n-1)(n-2)},$$

$$F^{**} = F - \frac{1}{n} FJ - \frac{1}{n} JF = F^{*},$$

$$A^{**\prime} = A^{**}, \qquad F^{**\prime} = -F^{**} \quad \text{and} \quad A^{**} + F^{**} = W^{**}$$

(Ω is defined above).

Evidently, we have analogous results in terms of D, B and G.

From [1] we now obtain the following important result for any n-matrix W, $n > 2$, which has all p.d.e. zero (each of A and F is such a matrix):

Theorem 1. *W has all row and column sums zero if and only if $W = W^{**}$.*

Corollary. *If W is real, if u and v are any $(n \times 1)$ real matrices, and if Λ is any real diagonal n-matrix, then*

$$\begin{aligned} \operatorname{tr}(A^{**})^2 - \operatorname{tr}(F^{**})^2 &= \operatorname{tr}(W^{**\prime} W^{**}) \\ &\leq \operatorname{tr}((W - uh' - hv' - \Lambda)'(W - uh' - hv' - \Lambda)), \end{aligned}$$

with equality on the right-hand side if and only if $W^{**} = W - uh' - hv' - \Lambda$.

Note. This corollary is not given explicitly in [1] but is easily obtained from results contained therein. The corollary, together with some equally easily obtained results, explains in some sense why the algorithm to be described shortly is, in general, more efficient than the original Lawler and Gilmore algorithms.

In [1], it is shown how the objective function for any given K.–B.P. always can be expressed in the form

$$K(P) = \operatorname{tr} P'(W'PD + C),$$

where W and D are fixed real n-matrices with all p.d.e. zero, C is a fixed real n-matrix, and P is a freely variable permutation matrix of order n, n being appropriately chosen.

The solution of the minimising variant of the K.–B.P. thus reduces to finding a permutation matrix $\check{P}$ of order n and $K(\check{P})$ where

$$K(\check{P}) = \operatorname{tr} \check{P}'(W'\check{P}D + C) = \min_{P} \operatorname{tr} P'(W'PD + C).$$

In [1] it was shown that, for each permutation matrix P of order $n > 2$,

$$K(\boldsymbol{P}) = \frac{1}{n}\operatorname{tr} \boldsymbol{J}\left(\boldsymbol{C} + \frac{\boldsymbol{AJB}}{n-1}\right)$$
$$+ \operatorname{tr} \boldsymbol{P}'\left(\boldsymbol{A}^{**}\boldsymbol{P}\boldsymbol{B}^{**} + \boldsymbol{F}^{**\prime}\boldsymbol{P}\boldsymbol{G}^{**} + \frac{2}{n-2}(\boldsymbol{AJB})^* + \frac{2}{n}(\boldsymbol{F}'\boldsymbol{JG})^* + \boldsymbol{C}^*\right),$$

this being the canonical form, referred to earlier, of the objective function of the K.–B.P.

As in [1], it is convenient to define

$$\boldsymbol{C}^{(1)*} = \frac{2}{n-2}(\boldsymbol{AJB})^* + \frac{2}{n}(\boldsymbol{F}'\boldsymbol{JG})^* + \boldsymbol{C}^*.$$

(Each of $(2/(n-2))(\boldsymbol{AJB})^*$, $(2/n)(\boldsymbol{F}'\boldsymbol{JG})^*$ and $\boldsymbol{C}^*$ has each row sum and each column sum zero and so the sum of these matrices has each row sum and each column sum zero; thus the use of the symbol $C^{(1)*}$ in this context is consistent with the definition of $\boldsymbol{X}^*$.) We now re-write equivalently the canonical form above as follows:

$$K(\boldsymbol{P}) = \frac{1}{n}\operatorname{tr} \boldsymbol{J}\left(\boldsymbol{C} + \frac{\boldsymbol{AJB}}{n-1}\right) + \operatorname{tr} \boldsymbol{P}'(\boldsymbol{A}^{**}\boldsymbol{P}\boldsymbol{B}^{**} + \boldsymbol{F}^{**\prime}\boldsymbol{P}\boldsymbol{G}^{**} + \boldsymbol{C}^{(1)*}). \tag{1}$$

We see that the solution of the K.–B.P. can be expressed equivalently as the determination of $\check{\boldsymbol{P}}$ and $K(\check{\boldsymbol{P}})$ where

$$K(\check{\boldsymbol{P}}) = \frac{1}{n}\operatorname{tr} \boldsymbol{J}\left(\boldsymbol{C} + \frac{\boldsymbol{AJB}}{n-1}\right) + \min_{\boldsymbol{P}} tr\, \boldsymbol{P}'(\boldsymbol{A}^{**}\boldsymbol{P}\boldsymbol{B}^{**} + \boldsymbol{F}^{**\prime}\boldsymbol{P}\boldsymbol{G}^{**} + \boldsymbol{C}^{(1)*}).$$

3. The algorithm

Any element of an n-matrix which is not a p.d.e. is called an off-diagonal element (o.d.e.) of that n-matrix.

Let $\mathscr{W}$ be the $((n-1)\times n)$ matrix formed from the o.d.e. of $\boldsymbol{W}$ in the following way:

Column j of $\mathscr{W}$ is formed of all $(n-1)$ o.d.e. of column j of $\boldsymbol{W}$ taken in such order that the elements of column j of $\mathscr{W}$ are non-increasing with increasing row number $(1 \le j \le n)$. $\mathscr{D}$, $\mathscr{A}$, $\mathscr{B}$, $\mathscr{F}$ and $\mathscr{G}$ are analogously formed from $\boldsymbol{D}$, $\boldsymbol{A}^{**}$, $\boldsymbol{B}^{**}$, $\boldsymbol{F}^{**}$ and $\boldsymbol{G}^{**}$, respectively.

$\check{\mathscr{D}}$ is the $((n-1)\times n)$ matrix formed from $\boldsymbol{D}$ as follows:

The elements of column j of $\check{\mathscr{D}}$ are formed from the $(n-1)$ o.d.e. of column j of $\boldsymbol{D}$ but, in this case, such that the elements of column j of $\check{\mathscr{D}}$ are non-decreasing with increasing row number $(1 \le j \le n)$; (equivalently, $\check{\mathscr{D}}$ can be formed from $\mathscr{D}$ by reversing the order of the elements within each column of $\mathscr{D}$). $\check{\mathscr{B}}$ and $\check{\mathscr{G}}$ are analogously formed from $\boldsymbol{B}^{**}$ and $\boldsymbol{G}^{**}$.

Now $\boldsymbol{P}'\boldsymbol{W}\boldsymbol{P}$ is a cogredient transformation of $\boldsymbol{W}$ for each permutation matrix $\boldsymbol{P}$ of order n, i.e. each o.d.e. of $\boldsymbol{P}'\boldsymbol{W}\boldsymbol{P}$ equals some o.d.e. of $\boldsymbol{W}$ and conversely, each p.d.e. of $\boldsymbol{P'WP}$ equals some p.d.e. of $\boldsymbol{W}$ and conversely, and each row of $\boldsymbol{P'W'P}$ $(=(\boldsymbol{P'WP})')$ is obtained by permuting the elements of some row of $\boldsymbol{W'}$, i.e. by permuting the elements of some column of $\boldsymbol{W}$; also, each row of $\boldsymbol{P'W'P}$ is some row of $\boldsymbol{W'P}$ without permutation of the elements of this row.

Since each p.d.e. of $\boldsymbol{W}$ and $\boldsymbol{D}$ is zero, it follows that $\operatorname{tr} \boldsymbol{P'W'PD}$ $(=\operatorname{tr}(\boldsymbol{P'WP})'\boldsymbol{D})$ is composed of $n(n-1)$ terms where each of these terms is the product of an o.d.e. of $\boldsymbol{W}$ and an o.d.e. of $\boldsymbol{D}$; each o.d.e. of both $\boldsymbol{W}$ and $\boldsymbol{D}$ occurs in precisely one of these $n(n-1)$ terms. Further, post-multiplication of $\boldsymbol{W'}$ by $\boldsymbol{P}$ at most permutes the elements of each row of $\boldsymbol{W'}$ but each element of row i of $\boldsymbol{W'P}$ in an element of row i of $\boldsymbol{W'}$ and conversely $(1 \le i \le n)$; also, the element in row i and column j of $\boldsymbol{W'PD}$ is the inner product of the row vector formed by row i of $\boldsymbol{W'P}$ and the column vector formed by column j of $\boldsymbol{D}$.

Since the elements of row i of $\mathcal{W}'$, the transpose of $\mathcal{W}$, are the o.d.e. from row i of $\boldsymbol{W'}$ in row i of $\boldsymbol{W'P}$, but taken in non-increasing order, and since the elements in column j of $\check{\mathfrak{D}}$ are the o.d.e. in column j of $\boldsymbol{D}$, but taken in non-decreasing order, and since we have seen that each o.d.e. of $\boldsymbol{W}$ and each o.d.e. of $\boldsymbol{D}$ occurs in precisely one of the $n(n-1)$ product terms referred to above as essentially forming $\operatorname{tr} \boldsymbol{P'W'PD}$, it follows that, if p_{ij} is a unit element of any permutation matrix $\boldsymbol{P}$, then the element in row i and column j of $\boldsymbol{W'PD} \ge$ the element in row i and column j of $\mathcal{W}'\check{\mathfrak{D}}$; by similar arguments, it follows that, if $p_{ij} = 1$, then the element in row i and column j of $\mathcal{W}'\mathfrak{D} \ge$ the element in row i and column j of $\boldsymbol{W'PD}$. It follows immediately

$$\min_{\boldsymbol{P}} \operatorname{tr} \boldsymbol{P}'(\mathcal{W}'\check{\mathfrak{D}} + \boldsymbol{C}) \le K(\check{\boldsymbol{P}}) = \min_{\boldsymbol{P}} \operatorname{tr} \boldsymbol{P}'(\boldsymbol{W'PD} + \boldsymbol{C})$$
$$\le \min_{\boldsymbol{P}} \operatorname{tr} \boldsymbol{P}'(\mathcal{W}'\mathfrak{D} + \boldsymbol{C}).$$

We note that each of

$$\min_{\boldsymbol{P}} \operatorname{tr} \boldsymbol{P}'(\mathcal{W}'\check{\mathfrak{D}} + \boldsymbol{C}) \quad \text{and} \quad \min_{\boldsymbol{P}} \operatorname{tr} \boldsymbol{P}'(\mathcal{W}'\mathfrak{D} + \boldsymbol{C})$$

is a linear assignment problem (perhaps in somewhat unfamiliar form); this is because $\mathcal{W}'\check{\mathfrak{D}}$ and $\mathcal{W}'\mathfrak{D}$ are fixed n-matrices since each of $\boldsymbol{W}$ and $\boldsymbol{D}$ is a fixed n-matrix; since $\boldsymbol{C}$ is a fixed n-matrix, it follows that each of the matrix sums $\mathcal{W}'\check{\mathfrak{D}} + \boldsymbol{C}$ and $\mathcal{W}'\mathfrak{D} + \boldsymbol{C}$ is a fixed n-matrix.

Also, we see that

$$\mathcal{W}'\check{\mathfrak{D}} + \boldsymbol{C} \le \mathcal{W}'\mathfrak{D} + \boldsymbol{C},$$

where the inequality is to be interpreted as holding between each pair of corresponding elements in the two matrix sums.

Now, the solution to a linear assignment problem is essentially routine, (see e.g. Kuhn [3]); moreover, since $\mathcal{W}'\check{\mathfrak{D}} + \boldsymbol{C} \le \mathcal{W}'\mathfrak{D} + \boldsymbol{C}$, it follows that any optimum dual feasible solution of the minimisation linear assignment problem with $\mathcal{W}'\check{\mathfrak{D}} + \boldsymbol{C}$

as cost matrix is a dual feasible solution to the corresponding linear assignment problem with $\mathcal{W}'\mathcal{D}+C$ as cost matrix; thus we can solve these two problems successively with less effort than the combined effort required to solve them both separately. Further, there is a heuristic basis for the belief that the permutation matrices, say P_1 and P_2, which give the minimum solutions to the two above primal linear assignment problems are "near", in some sense, to the permutation matrix $\check{P}$ where

$$K(\check{P}) = \min_P \operatorname{tr} P'(W'PD + C);$$

so it is sensible to evaluate the objective function $K(P)$ first putting $P = P_2$ and then putting $P = P_1$. We note that

$$\begin{aligned}\operatorname{tr} \check{P}'(W'\check{P}D + C) \le \operatorname{tr} P_2'(W'P_2D + C) &\le \operatorname{tr} P_2'(\mathcal{W}'\mathcal{D} + C)\\ &= \min_P \operatorname{tr} P'(\mathcal{W}'\mathcal{D} + C)\end{aligned}$$

and

$$\begin{aligned}\min_P \operatorname{tr} P'(\mathcal{W}'\check{\mathcal{D}} + C) = \operatorname{tr} P_1'(\mathcal{W}'\check{\mathcal{D}} + C) &\le \operatorname{tr} \check{P}'(W'\check{P}D + C)\\ &\le \operatorname{tr} P_1'(W'P_1D + C);\end{aligned}$$

finally, we see that

$$\operatorname{tr} P_1'(\mathcal{W}'\check{\mathcal{D}} + C) \le K(\check{P}) \le \min(\operatorname{tr} P_1'(W'P_1D + C), \operatorname{tr} P_2'(W'P_2D + C)).$$

Note. Essentially, the inequality: $\operatorname{tr} P_1'(\mathcal{W}'\mathcal{D} + C) \le K(\check{P})$ was one of the foundations of the Lawler and Gilmore algorithms for the K.–B.P. although each of these authors used a very different notation.

We now obtain bounds analogous to those described above to be used in conjunction with the canonical transformation of the objective function of the K.–B.P.; in this transformation implicitly we have linearised the objective function $K(P)$ as much as possible; this could be argued briefly as follows:

We recall the well-known inequality: if x and y are any real vectors with the same number of elements, then $|x'y| \le (x'x)^{1/2}(y'y)^{1/2}$. Slightly extending this inequality it is easy to see that

$$\begin{aligned}|\operatorname{tr} P'W'PD| &= |\operatorname{tr} P'APB + \operatorname{tr} P'F'PG|\\ &\le |\operatorname{tr} P'APB| + |\operatorname{tr} P'F'PG|\\ &\le (\operatorname{tr} A^2)^{1/2}(\operatorname{tr} B^2)^{1/2} + (-\operatorname{tr} F^2)^{1/2}(-\operatorname{tr} G^2)^{1/2}\\ &\le (\operatorname{tr} A^2 - \operatorname{tr} F^2)^{1/2}(\operatorname{tr} B^2 - \operatorname{tr} G^2)^{1/2}\\ &= (\operatorname{tr} W'W)^{1/2}(\operatorname{tr} D'D)^{1/2}\end{aligned}$$

and so

$$|\mathrm{tr}\, \boldsymbol{P}'\boldsymbol{W}'\boldsymbol{P}\boldsymbol{D}| \le (\mathrm{tr}\, \boldsymbol{A}^2)^{1/2}(\mathrm{tr}\, \boldsymbol{B}^2)^{1/2} + (-\mathrm{tr}\, \boldsymbol{F}^2)^{1/2}(-\mathrm{tr}\, \boldsymbol{G}^2)^{1/2}$$
$$\le (\mathrm{tr}\, \boldsymbol{W}'\boldsymbol{W})^{1/2}(\mathrm{tr}\, \boldsymbol{D}'\boldsymbol{D})^{1/2}.$$

Thereafter, writing $\boldsymbol{A}^{**}$, $\boldsymbol{B}^{**}$, $\boldsymbol{F}^{**}$, $\boldsymbol{G}^{**}$, $\boldsymbol{W}^{**}$ and $\boldsymbol{D}^{**}$ for $\boldsymbol{A}$, $\boldsymbol{B}$, $\boldsymbol{F}$, $\boldsymbol{G}$, $\boldsymbol{W}$ and $\boldsymbol{D}$, respectively, in the inequality immediately above and then using the corollary to Theorem 1 of Section 2, we see that

$$|\mathrm{tr}\, \boldsymbol{P}'\boldsymbol{A}^{**}\boldsymbol{P}\boldsymbol{B}^{**} + \mathrm{tr}\, \boldsymbol{P}'\boldsymbol{F}^{**\prime}\boldsymbol{P}\boldsymbol{G}^{**}| \le$$
$$\le |\mathrm{tr}\, \boldsymbol{P}'\boldsymbol{A}^{**}\boldsymbol{P}\boldsymbol{B}^{**}| + |\mathrm{tr}\, \boldsymbol{P}'\boldsymbol{F}^{**\prime}\boldsymbol{P}\boldsymbol{G}^{**}|$$
$$\le (\mathrm{tr}\, \boldsymbol{A}^{**2})^{1/2}(\mathrm{tr}\, \boldsymbol{B}^{**2})^{1/2} + (-\mathrm{tr}\, \boldsymbol{F}^{**2})^{1/2}(-\mathrm{tr}\, \boldsymbol{G}^{**2})^{1/2}$$
$$\le (\mathrm{tr}\, \boldsymbol{W}^{**\prime}\boldsymbol{W}^{**})^{1/2}(\mathrm{tr}\, \boldsymbol{D}^{**\prime}\boldsymbol{D}^{**})^{1/2}$$
$$\le (\mathrm{tr}\, \boldsymbol{W}'\boldsymbol{W})^{1/2}(\mathrm{tr}\, \boldsymbol{D}'\boldsymbol{D})^{1/2},$$

with

$$(\mathrm{tr}\, \boldsymbol{A}^{**2})^{1/2}(\mathrm{tr}\, \boldsymbol{B}^{**2})^{1/2} + (-\mathrm{tr}\, \boldsymbol{F}^{**2})^{1/2}(-\mathrm{tr}\, \boldsymbol{G}^{**2})^{1/2} =$$
$$= (\mathrm{tr}\, \boldsymbol{W}'\boldsymbol{W})^{1/2}(\mathrm{tr}\, \boldsymbol{D}'\boldsymbol{D})^{1/2}$$

if and only if either at least one of $\boldsymbol{W}$ and $\boldsymbol{D}$ is null or, at once, both $\boldsymbol{W} = \boldsymbol{W}^{**}$, $\boldsymbol{D} = \boldsymbol{D}^{**}$ and also $\mathrm{tr}\, \boldsymbol{A}^2\, \mathrm{tr}\, \boldsymbol{G}^2 = \mathrm{tr}\, \boldsymbol{B}^2\, \mathrm{tr}\, \boldsymbol{F}^2$.

Thus, we see that, by adopting the canonical form of the objective function of the K.–B.P., we have obtained a minimum upper bound for the absolute value of the quadratic terms in the objective function, i.e. of those terms, $\mathrm{tr}\, \boldsymbol{P}'\boldsymbol{W}'\boldsymbol{P}\boldsymbol{D}$ or $\mathrm{tr}\, \boldsymbol{P}'\boldsymbol{A}^{**}\boldsymbol{P}\boldsymbol{B}^{**} + \mathrm{tr}\, \boldsymbol{P}'\boldsymbol{F}^{**\prime}\boldsymbol{P}\boldsymbol{G}^{**}$, which cause the objective function $K(\boldsymbol{P})$ to adopt characteristics not possessed by the objective function of any linear assignment problem.

Using arguments directly analogous to those used earlier in this section, we see that

$$\min_{\boldsymbol{P}} \mathrm{tr}\, \boldsymbol{P}'(\mathscr{A}'\check{\mathscr{B}} + \mathscr{F}'\check{\mathscr{G}} + \boldsymbol{C}^{(1)*}) \le$$
$$\le \min_{\boldsymbol{P}} \mathrm{tr}\, \boldsymbol{P}'(\boldsymbol{W}'\boldsymbol{P}\boldsymbol{D} + \boldsymbol{C}) - \frac{1}{n}\, \mathrm{tr}\, \boldsymbol{J}\left(\boldsymbol{C} + \frac{\boldsymbol{A}\boldsymbol{J}\boldsymbol{B}}{n-1}\right)$$
$$\le \min_{\boldsymbol{P}} \mathrm{tr}\, \boldsymbol{P}'(\mathscr{A}'\mathscr{B} + \mathscr{F}'\mathscr{G} + \boldsymbol{C}^{(1)*}).$$

Since each of $\boldsymbol{A}^{**}$, $\boldsymbol{B}^{**}$, $\boldsymbol{F}^{**}$ and $\boldsymbol{G}^{**}$ has each column sum and each p.d.e. equal to zero, it follows that each column sum is zero in $\mathscr{A}$, $\mathscr{B}$, $\mathscr{F}$, $\mathscr{G}$, $\check{\mathscr{B}}$ and $\check{\mathscr{G}}$; thus

$$\mathscr{A}'\check{\mathscr{B}} + \mathscr{F}'\check{\mathscr{G}} \le \boldsymbol{o}_n \le \boldsymbol{O}_n \le \mathscr{A}'\mathscr{B} + \mathscr{F}'\mathscr{G}.$$

On this occasion we let $\boldsymbol{P}_1$, $\boldsymbol{P}_2$ and $\boldsymbol{P}_3$ be permutation matrices of order n giving optimum solutions to the three following linear assignment problems:

$$\min_{P} \operatorname{tr} P'(\mathscr{A}'\check{\mathscr{B}} + \mathscr{F}'\check{\mathscr{G}} + C^{(1)*}); \quad \min_{P} \operatorname{tr} P'C^{(1)*};$$

$$\min_{P} \operatorname{tr} P'(\mathscr{A}'\mathscr{B} + \mathscr{F}'\mathscr{G} + C^{(1)*}).$$

Since $\mathscr{A}'\check{\mathscr{B}} + \mathscr{F}'\check{\mathscr{G}} \leq O \leq \mathscr{A}'\mathscr{B} + \mathscr{F}'\mathscr{G}$, it follows that we can solve these three minimisation linear assignment problems successively with some saving of effort. Each of the permutation matrices P_1, P_2 and P_3 is likely to give a low value of $K(P)$ and thereafter, by arguments analogous to those used earlier in this section, we see that

$$\frac{1}{n} \operatorname{tr} J\left(C + \frac{AJB}{n-1}\right) + \operatorname{tr} P_1'(\mathscr{A}'\check{\mathscr{B}} + \mathscr{F}'\check{\mathscr{G}} + C^{(1)*}) \leq K(\check{P}) \leq \min_{P \in \{P_1, P_2, P_3\}} K(P).$$

Extending our earlier effective comparison of $|\operatorname{tr} P'W'PD|$ and $|\operatorname{tr} P'A^{**}PB^{**} + \operatorname{tr} P'F^{**\prime}PG^{**}|$ and to obtain some effective comparison between the linearisation $\operatorname{tr} P'\mathscr{W}'\check{\mathscr{D}}$ of $\operatorname{tr} P'W'PD$ and the linearisation $\operatorname{tr} P'(\mathscr{A}'\check{\mathscr{B}} + \mathscr{F}'\check{\mathscr{G}})$ of $\operatorname{tr} P'(A^{**}PB^{**} + F^{**\prime}PG^{**})$, we see first that, since all $n(n-1)$ elements of each of $\mathscr{W}$ and $\check{\mathscr{D}}$ are the $n(n-1)$ o.d.e. of W and D, respectively, and since W and D have all p.d.e. zero, it follows that

$$|\operatorname{tr} P'\mathscr{W}'\check{\mathscr{D}}| \leq (\operatorname{tr} P'\mathscr{W}'\mathscr{W}P)^{1/2}(\operatorname{tr} \check{\mathscr{D}}\check{\mathscr{D}}')^{1/2} = (\operatorname{tr} \mathscr{W}'\mathscr{W})^{1/2}(\operatorname{tr} \check{\mathscr{D}}\check{\mathscr{D}}')^{1/2} = (\operatorname{tr} W'W)^{1/2}(\operatorname{tr} D'D)^{1/2}.$$

Similarly,

$$|\operatorname{tr} P'\mathscr{W}'\mathscr{D}| \leq (\operatorname{tr} W'W)^{1/2}(\operatorname{tr} D'D)^{1/2}.$$

Next, we see that

$$\begin{aligned}
&|\operatorname{tr} P'(\mathscr{A}'\check{\mathscr{B}} + \mathscr{F}'\check{\mathscr{G}})| \leq \\
&\quad \leq |\operatorname{tr} P'\mathscr{A}'\check{\mathscr{B}}| + |\operatorname{tr} P'\mathscr{F}'\check{\mathscr{G}}| \\
&\quad \leq (\operatorname{tr} \mathscr{A}'\mathscr{A})^{1/2}(\operatorname{tr} \check{\mathscr{B}}'\check{\mathscr{B}})^{1/2} + (\operatorname{tr} \mathscr{F}'\mathscr{F})^{1/2}(\operatorname{tr} \check{\mathscr{G}}'\check{\mathscr{G}})^{1/2} \\
&\quad = (\operatorname{tr} A^{**2})^{1/2}(\operatorname{tr} B^{**2})^{1/2} + (-\operatorname{tr} F^{**2})^{1/2}(-\operatorname{tr} G^{**2})^{1/2} \\
&\quad \leq (\operatorname{tr} A^{**2} - \operatorname{tr} F^{**2})^{1/2}(\operatorname{tr} B^{**2} - \operatorname{tr} G^{**2})^{1/2} \\
&\quad = (\operatorname{tr} W^{**\prime}W^{**})^{1/2}(\operatorname{tr} D^{**\prime}D^{**})^{1/2} \\
&\quad \leq (\operatorname{tr} W'W)^{1/2}(\operatorname{tr} D'D)^{1/2},
\end{aligned}$$

with

$$\begin{aligned}
&(\operatorname{tr} A^{**2})^{1/2}(\operatorname{tr} B^{**2})^{1/2} + (-\operatorname{tr} F^{**2})^{1/2} - \operatorname{tr} G^{**2})^{1/2} = \\
&\quad = (\operatorname{tr} W'W)^{1/2}(\operatorname{tr} D'D)^{1/2}
\end{aligned}$$

if and only if either at least one of W and D is null or, at once, both $W = W^{**}$, $D = D^{**}$ and also $\operatorname{tr} A^2 \operatorname{tr} G^2 = \operatorname{tr} B^2 \operatorname{tr} F^2$.

Similarly,

$$|\text{tr}\, \boldsymbol{P}'(\mathscr{A}'\mathscr{B} + \mathscr{F}'\mathscr{G})| \leq |(\text{tr}\, \boldsymbol{A}^{**2})^{1/2}(\text{tr}\, \boldsymbol{B}^{**2})^{1/2} + (-\text{tr}\, \boldsymbol{F}^{**2})^{1/2}(-\text{tr}\, \boldsymbol{G}^{**2})^{1/2}$$

$$\leq (\text{tr}\, \boldsymbol{W}'\boldsymbol{W})^{1/2}(\text{tr}\, \boldsymbol{D}'\boldsymbol{D})^{1/2},$$

with equality on the right-hand side if and only if either at least one of $\boldsymbol{W}$ and $\boldsymbol{D}$ is null or, at once, both $\boldsymbol{W} = \boldsymbol{W}^{**}$, $\boldsymbol{D} = \boldsymbol{D}^{**}$ and also tr $\boldsymbol{A}^2$ tr $\boldsymbol{G}^2$ = tr $\boldsymbol{B}^2$ tr $\boldsymbol{F}^2$.

Thus we see that the absolute values of the linearised terms obtained from the residual quadratic term tr $\boldsymbol{P}'(\boldsymbol{A}^{**}\boldsymbol{P}\boldsymbol{B}^{**} + \boldsymbol{F}^{**\prime}\boldsymbol{P}\boldsymbol{G}^{**})$ in the canonical form of the objective function have upper bounds which never exceed, and are almost always less than, the corresponding upper bounds to the absolute values of the linearisations tr $\boldsymbol{P}'\mathscr{W}'\check{\mathscr{D}}$ and tr $\boldsymbol{P}'\check{\mathscr{W}}'\mathscr{D}$ of the quadratic term tr $\boldsymbol{P}'\boldsymbol{W}'\boldsymbol{P}\boldsymbol{D}$ in the original form of the objective function.

In addition to the methods of finding upper bounds for $K(\check{\boldsymbol{P}})$ described so far in this section, we also have the greedy heuristic described in [1]; often it will be useful to obtain this heuristic estimate prior to the carrying out of other steps within the algorithm; the canonical transformation of $K(\boldsymbol{P})$ is a necessary first stage both within the heuristic procedure and within the lower and upper bounding procedures described in this section.

We see that we have established methods for lower bounding and for upper bounding $K(\check{\boldsymbol{P}})$, the optimum value of the K.–B.P. objective function; we now need to establish a branching procedure for use within the branch and bound algorithm which we are in the process of constructing. The branching procedure is as follows:

We solve the lower bounding linear assignment problem:

$$\min_{\boldsymbol{P}} \text{tr}\, \boldsymbol{P}'(\mathscr{A}'\check{\mathscr{B}} + \mathscr{F}'\check{\mathscr{G}} + \boldsymbol{C}^{(1)*})$$

to find any optimum corresponding dual feasible solution for this problem together with at least one minimising permutation matrix $\boldsymbol{P}_1$. For each unit element p_{ij} of (each) $\boldsymbol{P}_1$ we find the corresponding dual variables u_i and v_j in the optimum dual feasible solution; for each such pair (i, j), as in Little et al.'s Travelling Salesman algorithm [5], we calculate θ_{ij}, the sum of the second smallest elements in row i and column j of the following (non-negative) matrix of residues:

$$\mathscr{A}'\check{\mathscr{B}} + \mathscr{F}'\check{\mathscr{G}} + \boldsymbol{C}^{(1)*} - \boldsymbol{u}\boldsymbol{h}' - \boldsymbol{h}\boldsymbol{v}',$$

where, as we recall from Section 2, $\boldsymbol{h}$ denotes the $(n \times 1)$ matrix of which each element is 1, where the $(n \times 1)$ matrix $\boldsymbol{u} = (u_1, u_2, \ldots, u_n)$ and where the $(1 \times n)$ matrix $\boldsymbol{v}' = [v_1, v_2, \ldots, v_n]$.

Thereafter: we find

$$\hat{\theta} = \theta_{i^*, j^*} = \max_{(i,j)}\{\theta_{ij} : p_{ij} = 1\},$$

i.e. where we take the maximum over all unit elements p_{ij} of $\boldsymbol{P}_1$ and where p_{i^*,j^*} is such an element.

Our branching procedure consists in putting $p_{i^*,j^*}=1$ in the variable permutation matrix $\boldsymbol{P}$ within $K(\boldsymbol{P})$, our K.–B.P. objective function. In [1] it is shown in detail how, conditional upon putting equal to 1 any element of $\boldsymbol{P}$, say $p_{i^*,j^*}=1$ (but otherwise leaving $\boldsymbol{P}$ as a freely varying permutation matrix of order n), the canonical form of the objective function of our K.–B.P. of order n is reduced to a derived but otherwise unconstrained K.–B.P. of order $(n-1)$ where this smaller K.–B.P. objective function is already in canonical form (we note that, if $p_{i^*,j^*}=1$ is an element of $\boldsymbol{P}$, a permutation matrix of order n, and if row i^* and column j^* are removed from $\boldsymbol{P}$, a permutation matrix of order $(n-1)$ remains and this smaller permutation matrix is wholly unconstrained if $\boldsymbol{P}$ is only constrained such that $p_{i^*,j^*}=1$). The derived K.–B.P. of order $(n-1)$ is next labelled with provisional commitment that $p_{i^*,j^*}=1$; this smaller problem will be lower bounded in its turn using the procedure that we have described in detail above for the K.–B.P. of order n.

In order to find the lower bound of $K(\boldsymbol{P})$ for all permutation matrices $\boldsymbol{P}=[p_{ij}]$ where $p_{i^*,j^*}\neq 1$, we replace by the very large positive number M the element in row i^* and column j^* of our matrix

$$\mathcal{A}\check{\mathcal{B}}+\mathcal{F}'\check{\mathcal{G}}+\boldsymbol{C}^{(1)*}$$

and then, using $\boldsymbol{u}$ and $\boldsymbol{v}'$ as the starting dual feasible solution row and column matrices, respectively, we obtain the maximum dual feasible solution (and thereby the minimum primal solution) to the linear assignment problem with the above modified matrix as cost matrix; evidently, if $u_1^*, u_2^*, \ldots, u_n^*, v_1^*, v_2^*, \ldots, v_n^*$ are optimum dual variables for this new linear assignment problem, then

$$\sum_{i=1}^{n}(u_i^*+v_i^*-u_i-v_i)\geq\theta_{i^*,j^*}.$$

The required lower bound of $K(\boldsymbol{P})$, for all permutation matrices $\boldsymbol{P}$ where $p_{i^*,j^*}\neq 1$, is

$$\sum_{i=1}^{n}(u_i^*+v_i^*)+\frac{1}{n}\operatorname{tr}\boldsymbol{J}'\left(\boldsymbol{C}+\frac{\boldsymbol{AJB}}{n-1}\right).$$

In this section we have described how, during a single stage of the algorithm to determine $K(\check{\boldsymbol{P}})$ and $\check{\boldsymbol{P}}$, we bound $K(\check{\boldsymbol{P}})$ both from below and from above, determining a permutation matrix for which the objective function achieves the lowest upper bound for $K(\check{\boldsymbol{P}})$; we have outlined how, in the canonical form of $K(\boldsymbol{P})$, the freely variable permutation matrix $\boldsymbol{P}$ can be constrained to have some element $p_{i^*,j^*}=1$ and so, thereafter, to leave a K.–B.P. of order $(n-1)$ which is not constrained and is already in canonical form. Also, we have shown how the lower bound of $K(\boldsymbol{P})$ can be obtained for all $\boldsymbol{P}=[p_{ij}]$ where $p_{i^*,j^*}\neq 1$; further, it is easy to see how the constraint $p_{i^*,j^*}\neq 1$, and all required similar constraints, can be retained by the continued inclusion of a large positive number M in place of the original element in row i^* and column j^* and in place of each other original element corresponding to a "prohibited" row and column pair.

As indicated above, we refer the reader to [1] for the complete description of the procedure by which the canonical form of $K(\boldsymbol{P})$, of order n, is transformed into the canonical form of a K.–B.P. of order $(n-1)$ whenever any single element of $\boldsymbol{P}$ is constrained to be 1 (the detailed re-statement here of this transformation would require a considerable amount of further carefully defined notation).

If we keep a record of each element of $\boldsymbol{P}$ which has been given unit value, (followed by the removal of the row and column of $\boldsymbol{P}$ containing this element and the modification of the canonical form of the objective function), then, using combinations and iterations of the steps described hitherto in this section, it is clear that we can construct a branch and bound algorithm for the K.–B.P.; in the next section we give an example of the application of such an algorithm to a specific K.–B.P.

4. An example of the algorithm

To illustrate the method of solution proposed and described in the previous section, we take an example solved by Lawler [4]; in this example both $\boldsymbol{W}$ and $\boldsymbol{D}$ are symmetric and so $\boldsymbol{A}=\boldsymbol{W}$, $\boldsymbol{F}=\boldsymbol{0}$, $\boldsymbol{B}=\boldsymbol{D}$ and $\boldsymbol{G}=\boldsymbol{0}$. So

$$\boldsymbol{A}=\boldsymbol{W}=\begin{bmatrix} 0&0&6&1&1&8&4\\ 0&0&1&0&3&1&3\\ 6&1&0&8&8&4&2\\ 1&0&8&0&7&6&4\\ 1&3&8&7&0&0&6\\ 8&1&4&6&0&0&9\\ 4&3&2&4&6&9&0 \end{bmatrix}; \qquad \boldsymbol{B}=\boldsymbol{D}=\begin{bmatrix} 0&5&0&5&0&5&4\\ 5&0&9&7&3&8&6\\ 0&9&0&9&4&4&4\\ 5&7&9&0&1&1&9\\ 0&3&4&1&0&5&5\\ 5&8&4&1&5&0&4\\ 4&6&4&9&5&4&0 \end{bmatrix};$$

$$\boldsymbol{C}=\begin{bmatrix} 51&27&14&9&0&18&0\\ 0&1&22&17&0&41&13\\ 2&0&13&22&2&12&27\\ 38&11&0&0&22&13&14\\ 62&56&0&67&1&0&5\\ 61&0&3&14&9&1&67\\ 41&12&23&0&18&41&0 \end{bmatrix}.$$

Using the definitions for $\boldsymbol{A}^{**}$, $\mathscr{A}$ and $\mathscr{B}$ given in Sections 2 and 3, we see

$$A^{**} = \begin{bmatrix} 0 & -0.13 & 1.66 & -2.74 & -2.53 & 3.87 & -0.13 \\ -0.13 & 0 & -0.94 & -1.34 & 1.87 & -0.73 & 1.27 \\ 1.66 & -0.94 & 0 & 2.47 & 2.67 & -1.93 & -3.93 \\ -2.74 & -1.34 & 2.47 & 0 & 2.27 & 0.67 & -1.33 \\ -2.53 & 1.87 & 2.67 & 2.27 & 0 & -5.14 & 0.86 \\ 3.87 & -0.73 & -1.93 & 0.67 & -5.14 & 0 & 3.26 \\ -0.13 & 1.27 & -3.93 & -1.33 & 0.86 & 3.26 & 0 \end{bmatrix};$$

$$\mathscr{A} = \begin{bmatrix} 3.87 & 1.87 & 2.67 & 2.47 & 2.67 & 3.87 & 3.26 \\ 1.66 & 1.27 & 2.47 & 2.27 & 2.27 & 3.26 & 1.27 \\ -0.13 & -0.13 & 1.66 & 0.67 & 1.87 & 0.67 & 0.86 \\ -0.13 & -0.73 & -0.94 & -1.33 & 0.86 & -0.73 & -0.13 \\ -2.53 & -0.94 & -1.93 & -1.34 & -2.53 & -1.93 & -1.33 \\ -2.74 & -1.34 & -3.93 & -2.74 & -5.14 & -5.14 & -3.93 \end{bmatrix};$$

$$B^{**} = \begin{bmatrix} 0 & 0.13 & -3.26 & 1.33 & -0.86 & 2.33 & 0.33 \\ 0.13 & 0 & 1.93 & -0.46 & -1.66 & 1.53 & -1.47 \\ -3.26 & 1.93 & 0 & 3.13 & 0.93 & -0.86 & -1.87 \\ 1.33 & -0.46 & 3.13 & 0 & -2.46 & -4.27 & 2.73 \\ -0.86 & -1.66 & 0.93 & -2.46 & 0 & 2.52 & 1.53 \\ 2.33 & 1.53 & -0.86 & -4.27 & 2.52 & 0 & -1.25 \\ 0.33 & -1.47 & -1.87 & 2.73 & 1.53 & -1.25 & 0 \end{bmatrix};$$

$$\check{\mathscr{B}} = \begin{bmatrix} -3.26 & -1.66 & -3.26 & -4.27 & -2.46 & -4.27 & -1.87 \\ -0.86 & -1.47 & -1.87 & -2.46 & -1.66 & -1.25 & -1.47 \\ 0.13 & -0.46 & -0.86 & -0.46 & -0.86 & -0.86 & -1.25 \\ 0.33 & 0.13 & 0.93 & 1.33 & 0.93 & 1.53 & 0.33 \\ 1.33 & 1.53 & 1.93 & 2.73 & 1.53 & 2.33 & 1.53 \\ 2.33 & 1.93 & 3.13 & 3.13 & 2.52 & 2.52 & 2.73 \end{bmatrix};$$

$$\mathscr{A}'\check{\mathscr{B}} = -\begin{bmatrix} 23.85 & 17.98 & 29.19 & 36.20 & 23.06 & 31.49 & 20.91 \\ 11.82 & 9.03 & 15.05 & 18.78 & 12.09 & 16.14 & 10.54 \\ 22.65 & 19.49 & 31.65 & 37.06 & 25.83 & 31.75 & 24.69 \\ 18.52 & 15.26 & 25.27 & 30.44 & 20.61 & 26.02 & 18.76 \\ 26.03 & 22.31 & 34.73 & 39.70 & 27.97 & 33.38 & 28.27 \\ 30.12 & 24.49 & 39.78 & 47.18 & 32.09 & 39.74 & 30.09 \\ 22.58 & 17.31 & 28.73 & 33.54 & 22.92 & 29.45 & 21.84 \end{bmatrix}.$$

(We compare the above matrix $\mathcal{A}'\check{\mathcal{B}}$ with the analogous matrix of quadratic term approximation, produced by Lawler's original method of solution, $\mathcal{W}'\check{\mathcal{D}}$, as follows:

$$\mathcal{W}'\check{\mathcal{D}} = \begin{bmatrix} 26 & 93 & 53 & 50 & 27 & 58 & 83 \\ 9 & 37 & 20 & 18 & 10 & 24 & 33 \\ 59 & 153 & 99 & 101 & 57 & 102 & 129 \\ 49 & 131 & 77 & 82 & 46 & 84 & 110 \\ 44 & 124 & 73 & 75 & 42 & 80 & 105 \\ 49 & 139 & 81 & 94 & 47 & 90 & 118 \\ 61 & 151 & 101 & 108 & 59 & 100 & 132 \end{bmatrix}$$

We note that $\boldsymbol{W}$ and $\boldsymbol{D}$ are both quite sparse and non-negative but that, nevertheless, the sum of the smallest (largest in absolute value) elements in each row of $\mathcal{A}'\check{\mathcal{B}}$ is -242.9 whilst the sum of the smallest elements in each row of $\mathcal{W}'\check{\mathcal{D}}$ is 286.)

Using the definition of $\boldsymbol{X}^*$ for any n-matrix $\boldsymbol{X}$, since $\boldsymbol{A}h = (20, 8, 29, 26, 25, 28, 28)$ and $\boldsymbol{h}'\boldsymbol{B} = [19, 38, 30, 32, 18, 27, 32]$, and since $n = 7$ in our example, it follows that

$$\frac{2}{n-2}(\boldsymbol{AJB})^* =$$

$$= \begin{bmatrix} 12.34 & -13.72 & -2.74 & -5.49 & 13.72 & 1.37 & -5.49 \\ 55.54 & -61.72 & -12.34 & -24.69 & 61.72 & 6.17 & -24.69 \\ -20.06 & 22.28 & 4.46 & 8.91 & -22.28 & -2.29 & 8.91 \\ -9.26 & 10.28 & 2.06 & 4.11 & -10.28 & -1.02 & 4.11 \\ -5.66 & 6.28 & 1.26 & 2.51 & -6.28 & -0.63 & 2.51 \\ -16.46 & 18.28 & 3.66 & 7.31 & -18.28 & -1.83 & 7.31 \\ -16.46 & 18.28 & 3.66 & 7.31 & -18.28 & -1.83 & 7.31 \end{bmatrix};$$

$$\boldsymbol{C}^* = \begin{bmatrix} 15.33 & 12.47 & 4.04 & -8.67 & -6.67 & 0.76 & -17.24 \\ -32.10 & -9.96 & 15.61 & 2.90 & -3.10 & 27.33 & -0.67 \\ -27.82 & -8.67 & 8.90 & 10.18 & 1.18 & 0.61 & 15.61 \\ 5.33 & -0.53 & -6.96 & -14.67 & 18.33 & -1.24 & -0.24 \\ 16.04 & 31.18 & -20.24 & 39.04 & -15.96 & -27.53 & -22.53 \\ 20.18 & -19.67 & -12.10 & -8.82 & -2.82 & -21.39 & 44.61 \\ 3.08 & -4.82 & 10.76 & -19.96 & 9.04 & 21.47 & -19.53 \end{bmatrix};$$

$$\frac{2}{n-2}(\boldsymbol{AJB})^* + \boldsymbol{C}^* =$$

$$= \begin{bmatrix} 27.67 & -1.25 & 1.30 & -14.16 & 7.05 & 2.13 & -22.73 \\ 23.44 & -71.68 & 3.27 & -21.79 & 58.62 & 33.50 & -25.36 \\ -47.88 & 13.61 & 13.36 & 19.09 & -21.10 & -1.62 & 24.52 \\ -3.93 & 9.75 & -4.90 & -10.56 & 8.05 & -2.27 & 3.87 \\ 10.38 & 37.46 & -18.98 & 41.55 & -22.24 & -28.16 & -20.02 \\ 3.72 & -1.39 & -8.44 & -1.51 & -21.10 & -23.22 & 51.92 \\ -13.38 & 13.46 & 14.42 & -12.65 & -9.24 & 19.64 & -12.22 \end{bmatrix}.$$

(From [1] we see that the expected value of the K.–B.P. objective function, where the expectation is taken as $\boldsymbol{P}$ varies uniformly over all 7! permutation matrices of order 7, is

$$\tfrac{1}{7}\operatorname{tr}\boldsymbol{J}\left(\boldsymbol{C}+\frac{\boldsymbol{AJB}}{7-1}\right) = 124\tfrac{2}{7} + 765\tfrac{1}{3} = 889.62;$$

we see that the smallest element in $\frac{2}{5}(\boldsymbol{AJB})^* + \boldsymbol{C}^*$ is -71.68 and that this is the element in row 2 and column 2 of this latter matrix sum. As shown in [1], the expected value of the K.–B.P. objective function, as $\boldsymbol{P}$ varies uniformly over all 6! permutation matrices of order 7 in which $p_{22} = 1$, is

$$889.62 - \tfrac{7}{6}(71.68) = 805.99.$$

$$\mathscr{A}'\check{\mathscr{B}} + \tfrac{2}{5}(\boldsymbol{AJB})^* + \boldsymbol{C}^* =$$

$$= \begin{bmatrix} 3.82 & -19.23 & -27.89 & -50.36 & -16.01 & -29.36 & -43.64 \\ 11.62 & -80.71 & -11.78 & -40.57 & 46.53 & 17.36 & -35.90 \\ -70.53 & -5.88 & -18.29 & -17.97 & -46.93 & -33.37 & -0.17 \\ -29.96 & -12.56 & -30.17 & -41.00 & -12.56 & -28.29 & -14.89 \\ -15.65 & 15.15 & -53.71 & 1.85 & -50.21 & -61.54 & -48.29 \\ -26.40 & -25.88 & -48.22 & -48.69 & -53.19 & -62.96 & 21.83 \\ -35.96 & -3.85 & -14.31 & -46.19 & -32.16 & -9.81 & -34.06 \end{bmatrix}.$$

As stated earlier, we denote by $\boldsymbol{P}_1$ the permutation matrix of order 7 corresponding to the minimum primal solution to the linear assignment problem with the above matrix as cost matrix; the minimum value of the objective function of this linear assignment problem is -384.71, which when added to $\frac{1}{7}\operatorname{tr}\boldsymbol{J}(\boldsymbol{C}+\frac{1}{6}\boldsymbol{AJB}) = 889.62$, gives 504.91 (or 505 as an integer) as the first lower bound for $K(\check{\boldsymbol{P}})$, the minimum value of the K.–B.P. objective function.

Now, it is convenient to denote each permutation matrix of order n by a

$(1 \times n)$ matrix, where the element in column j is the number of the unique row where the corresponding permutation matrix contains a unit element in *its* column j; the elements of each such $(1 \times n)$ matrix are a permutation of the integers $1, 2, \ldots, n$; each such $(1 \times n)$ matrix denotes in this way a unique permutation matrix of order n and conversely. It is easy to show that, corresponding to the permutation matrix $\boldsymbol{P}_1$ referred to in the preceding paragraph, we have the (1×7) matrix $[3, 2, 5, 4, 7, 6, 1]$. This means of denoting a permutation matrix of order 7 will be adopted in the tree diagram for the solution of our particular problem example; further, in the diagram, if such a (1×7) matrix is preceded by a figure >500, then the figure will denote a lower bound of some subset of feasible solutions of the K.–B.P., and the (1×7) matrix will indicate the permutation matrix associated with that bound in the same way that $\boldsymbol{P}_1$ is associated with the lower bound 505 for all feasible solutions of our problem; if a figure >500 follows one of these (1×7) matrices, then this figure is $K(\boldsymbol{P})$ for the particular permutation matrix $\boldsymbol{P}$ corresponding to the (1×7) matrix. (The association of a "policy", used in a sense similar to that used in the context of dynamic programming, with each lower bound and the "re-injection" of that policy into the objective function to find an upper bound for the minimum solution value are characteristics of the algorithm described here.)

Of the unit elements in $\boldsymbol{P}_1$, that in row 2 and column 2 gives a penalty cost of 69 to be added to the lower bound at 505 for all permutation matrices $\boldsymbol{P} = [p_{ij}]$ such that $p_{22} \neq 1$; moreover, the greedy approximator described in [1] and of which we described the first stage two or so pages ago (to give an expected value of $K(\boldsymbol{P})$ of 805.99 for $\boldsymbol{P}$ such that $p_{22} = 1$) gives the permutation matrix corresponding to [3, 2, 4, 7, 5, 6, 1] with $K(\boldsymbol{P}) = 559$ for this $\boldsymbol{P}$. It follows that we know that each minimising $\boldsymbol{P}$ is such that $p_{22} = 1$.

In [1] we have described in considerable detail how, for example when $p_{22} = 1$, the canonical form of the objective function of our K.–B.P. can be transformed into the canonical form of a derived K.–B.P. of order $n - 1 = 6$; whereupon all the steps described in this section can be repeated to find a new lower bound, not less than the bound of 505 already found, and an associated permutation matrix of order 6 which, together with the element $p_{22} = 1$, gives a permutation matrix of order 7 (in which $p_{22} = 1$) which in its turn gives a new candidate upper bound to the minimum objective function. The algorithm then proceeds in similar fashion with however many steps and backtracks as are required until the usual branch and bound conditions for optimality are satisfied. In our particular problem, 559 is the minimum value of the objective function and $\check{\boldsymbol{P}}$ is the permutation matrix corresponding, in the manner described earlier, to the (1×7) matrix [3, 2, 4, 7, 5, 6, 1]; (this is also the "approximate" solution determined by the greedy approximator).

The progress of the algorithm to solve our example is illustrated by the solution tree which embodies the notation and conventions described above (Fig. 1).

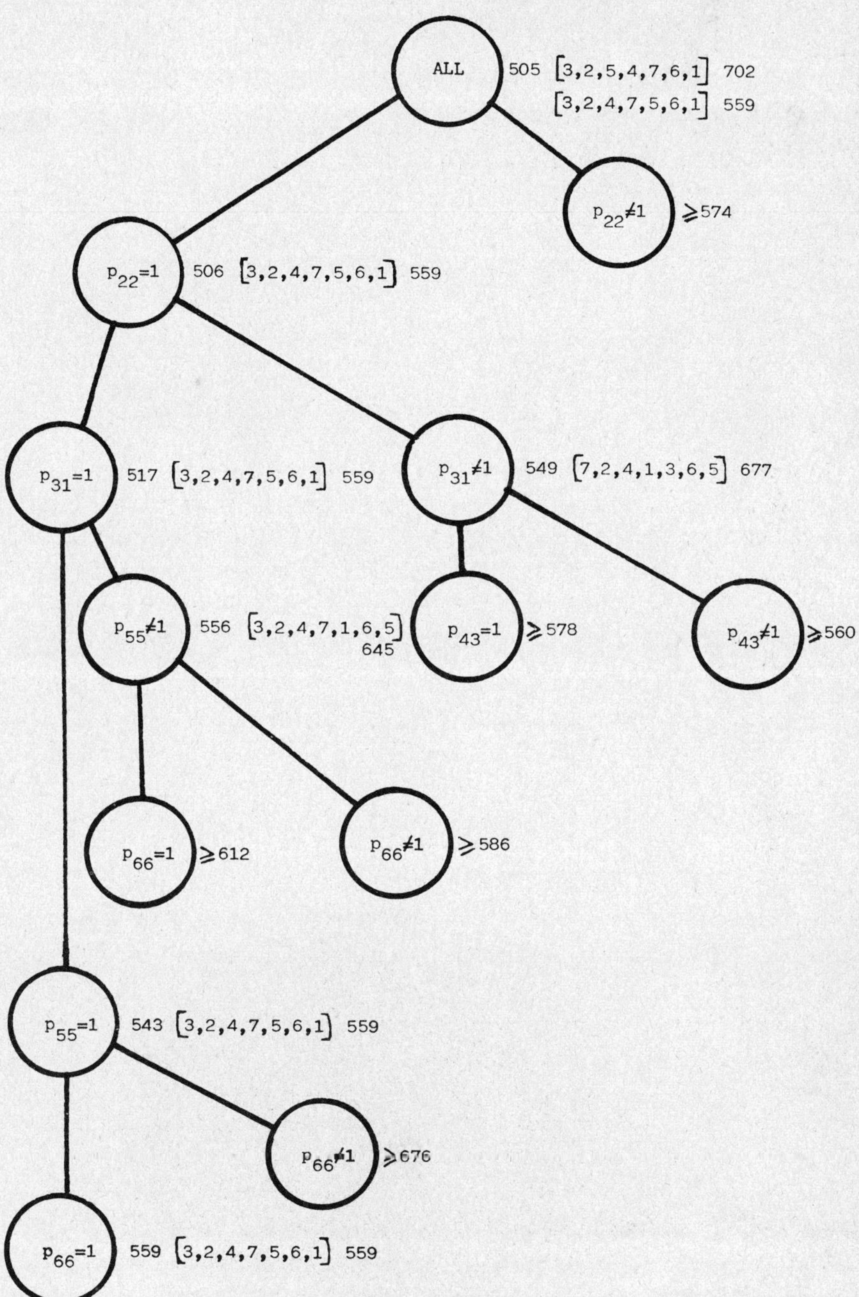

Fig. 1. An optimum solution with $K(\check{P}) = 559$. Inspection of the above tree shows that any minimising permutation matrix $\check{P}$ is of the form $[3, 2, -, -, 5, 6, -]$.

References

[1] C.S. Edwards, "The derivation of a greedy approximator for the Koopmans–Beckman quadratic assignment problem", in: T.B. Boffey, ed., *Proceedings of the CP77 Combinatorial Programming Conference* (Liverpool University, 1977) pp. 55–86.

[2] P.C. Gilmore, "Optimal and suboptimal algorithms for the quadratic assignment problem", *Journal of the Society for Industrial and Applied Mathematics* 10 (1962) 305–313.

[3] H.W. Kuhn, "The Hungarian method for the assignment problem", *Naval Research Logistic Quarterly* 2 (1955) 83–97.

[4] E.L. Lawler, "The quadratic assignment problem", *Management Science* 9 (1963) 586–599.

[5] J.D.C. Little, K.G. Murty, D.W. Sweeney and C. Karel, "An algorithm for the travelling salesman problem", *Operations Research* 11 (1963) 972–989.

Mathematical Programming Study 13 (1980) 53–57.
North-Holland Publishing Company

A PROBLEM OF SCHEDULING CONFERENCE ACCOMMODATION

A.I. HINXMAN

Edinburgh Regional Computing Centre, Edinburgh, Scotland

Received 1 February 1980

A University hall of residence consists of a number of buildings, or houses, which are used during vacations to accommodate the delegates to conferences held at the University. For brevity, the totality of delegates attending a conference will be referred to as the conference.

As conference bookings are made, the conferences are assigned to the houses in which they will be accommodated. The problem studied in this paper is that of keeping to a minimum for each conference the number of different houses in which delegates of that conference are accommodated.

The model adopted is one in which all the bookings for the period under consideration are known at the start of the period and the problem is to make the assignments of conferences to accommodation in such a way as to maximuse the utility under the compactness criterion.

Key words: Algorithm Analysis, Best-fit, Booking, Branch-and-backtrack, Compactness, Conference, Heuristics, Optimization, State-space Search, Utilization.

1. Introduction

A university hall of residence consists of a number of buildings, *houses*, which are used during vacations to accommodate organised groups, such as delegates to conferences. For simplicity, the totality of members of such a group will be referred to as a *conference*. A period of time in which the hall is in use in this way is called a *booking period*.

As conference bookings are made, the conferences are assigned to the houses in which they will be accommodated. The administration of the hall seeks to maximuse the utility of these assignments according to a number of criteria. These include:

(i) *utilisation*, as much of the accommodation as possible should be in use,

(ii) *compactness*, the number of different houses in which members of a particular conference are accommodated should be as small as possible,

(iii) *acceptability*, some conferences may require accommodation with specified facilities, e.g. double rooms, rooms with wash-hand basins,

(iv) *stability*, initially the assignments are tentative, and can be altered in the light of bookings and cancellations, but once literature including accommodation details has been printed it is highly undesirable to make any further alterations to the relevant assignments.

2. A simple model

It is a sufficiently complicated problem to consider simply how compactness may be achieved.

Let the houses be numbered $i = 1, 2, \ldots, n$, the days of the booking period be numbered $j = 1, 2, \ldots, p$, and the conferences be numbered $k = 1, 2, \ldots, q$.

Let h_i be the size of the ith house, s_k be the size of the kth conference, d_k be the day of arrival of the kth conference, and r_k be the length of stay of the kth conference.

Let c_{ijk} be the number of members of conference k accommodated in house i on day j.

Let

$$z_{ik} = \begin{cases} 0 & \text{if } \sum_j c_{ijk} = 0, \\ 1 & \text{otherwise.} \end{cases}$$

Then it is required to

$$\text{minimise} \quad \sum_{i,k} z_{ik},$$

$$\text{subject to} \quad c_{ijk} \geq 0 \quad \text{for all } i, j, k;$$

$$\sum_i c_{ijk} = \begin{cases} s_k & \text{for all } j \text{ such that } d_k \leq j < d_k + r_k, \\ 0 & \text{otherwise;} \end{cases}$$

$$c_{ijk} = c_{i(j-1)k} \quad \text{for any } j \text{ such that } d_k < j < d_k + r_k;$$

$$\sum_k c_{ijk} \leq h_i \quad \text{for all } j.$$

The non-zero c_{ijk}'s for a given k and a j in the range $d_k \leq j < d_k + r_k$ are the sizes of the *components* of conference k. The formulation given does not control the relative sizes of the components. Neither does it control the number of (i, j) pairs for which $\sum_k c_{ijk} \neq 0$. ($\sum_k c_{ijk} = 0$ means that house i has no occupants on day j.) In circumstances where these values are significant, extensions to the model would be needed.

3. Evaluation of heuristics

In the absence of an algorithm for solution of the stated problem, solution methods will be heuristic in nature. Some measure must be made of the adequacy of an heuristic solution.

The remarks that follow are appropriate to *incremental* solution methods. By an incremental solution method is meant one in which the sequence of events is:

(1) a conference that has not yet been allocated accommodation is selected from the list of conferences,

(2) an allocation of accommodation is made for this conference,

(3) adjustments are possibly made to the allocations for this and previously allocated conferences to produce a partial solution that is "better" overall,

(4) return to step (1) to allocate accommodation for a further conference.

For conference k, let $t(k)$ be the smallest number of components it could have if no other conference were in residence during the time k is in residence. Clearly the actual number of components k has in a solution may be greater than $t(k)$; consider for example two houses of size 200 and 100 respectively and two contemporaneously resident conferences each of size 150.

Suppose that in an incremental solution method, P, the conferences are allocated accommodation in the order in which they are numbered. Let $a_2(l, P, k)$ be the number of components that conference l has at the end of step 2 of iteration k of method P. If $a_2(k, P, k) - t(k) \neq 0$, then its value is a discrepancy which either should be justified or indicates that some action should be taken at step 3.

Let $t_l(k, P)$ be the smallest number of components k can have if the values of $a_2(l, P, k)$, $l < k$ are regarded as fixed. In practice it may be too difficult to calculate $t_l(k, P)$. Let $t_2(k, P)$ be a function that almost always takes the same value as $t_l(k, P)$ and never takes a value that is greatly different. If $a_2(k, P, k) = t_2(t, p)$, the discrepancy can be regarded as justified.

Otherwise attempts are made to reduce the discrepancy. Let $a_3(l, P, k)$ be the number of components that conference l has at the end of step 3 of iteration k of method P. If P is such that $a_3(l, P, k) = a_2(l, P, k)$ for all $l < k$, then $a_3(k, P, k) - t_2(k, P)$ is identified as the discrepancy for conference k that has neither been justified nor eliminated. In any case

$$\sum_{k=1}^{q} [a_3(k, P, q) - t_2(k, P)],$$

the *badness of fit* of P for the given data set, is a measure of the amount of splitting of conferences into components that can neither be justified nor eliminated.

4. Results from an heuristic method

Some investigation of an heuristic method has been done using house sizes of 341, 215, 193, 158, 156, 154, 154, 154 and generated booking information (Hastings and Peacock [1]) based on observed data. No assumptions were made as to the relationship between date of arrival and length of stay, as whilst academic conferences might not be expected to overstay a weekend, an opera company might rehearse over a week-end.

The method used involved a best-fit allocation (Knuth [2]) for step 2 and state-space search (Nilsson [3]) and branch-and-backtrack (Scott [4]) in step 3.

Table 1
Results from an heuristic program

Data set	Number of conferences	Percentage occupancy	Execution time(secs)	Badness of fit	Comments
1	169	70.51	465	15	
2	181	69.08	326	5	
3	158	65.46	48	5	
4	149	63.46	122	11	
5	191	78.47	954	29	
6	176	72.74	244	16	
7	166	51.92	150	10	
8	172	72.06	170	14	1
9	162	55.19	162	10	
10	162	58.61	31	1	
11	184	57.32	31	5	
12	165	52.95	45	3	
13	178	68.43	71	1	
14	190	70.44	1417	20	2
15	180	72.60	270	21	
16	167	52.43	114	8	
17	173	53.72	87	8	

Comments

(1) A plan with badness of fit 11 was generated during the development of the program.

(2) An incomplete plan with badness of fit 3 for the first 82 conferences was generated during development of the program. The plan reported here has badness of fit 4 for the first 82 conferences.

The order of allocation of the conferences was determined by the rules:

(i) if $d_k < d_l$, then $k < l$,
(ii) if $d_k = d_l$ and $s_k > s_l$, then $k < l$,
(iii) if $d_k = d_l$ and $s_k = s_l$ and $r_k > r_l$, then $k < l$.

The reason for rule (ii) is that it is generally more difficult to make satisfactory allocations for large conferences than for small ones. Rule (iii) arises from the fact that if two conferences of equal size arriving on the same day are allocated accommodation, the first allocation will usually be "better" and is therefore the one that should persist longer.

It is a simplification to say that it is more difficult to make satisfactory allocations for large conferences than small ones. For example, with the house sizes given there is only one allocation with the minimum possible number of components for a conference of size 340, whilst for a conference of size 342 there are three essentially different ones ($341 + 1$, $215 + 127$,$193 + 149$). So if size of conference were the primary criterion for the order in which they were allocated accommodation, some measure of relative "awkwardness" of different sizes would have to be developed.

There were two reasons in the present work for choosing to order on day or arrival (rule (i) above). Firstly, in a practical application bookings for more distant dates are more volatile, and therefore extra splitting of such conferences is more tolerable. Secondly, the logic of both the justification and adjustment steps of an incremental solution method is more straightforward if only conferences arriving no later than the conference being allocated accommodation have to be considered.

The method was implemented in IMP (Stephens [5]) to give the results shown in Table 1. During the development of an heuristic program a large number of total and partial solutions to problems are generated. In this case these solutions suggest that except for data sets 9 and 14 (see comments at foot of table) and badness of fit results from inadequacies in the justification part of the program.

Acknowledgment

This work was supported in part by SRC grant B/RG/83176. The grant holder was Dr T. B. Boffey, to whom the author is grateful for his encouragement. The referees made many helpful comments on an earlier version of this paper. Much of the computing involved has been done on the equipment of the E.R.C.C., to whom thanks are also due.

References

[1] N.A.J. Hastings and J.B. Peacock, *Statistical distributions* (Bullerworths, London, 1974).

[2] D.E. Knuth, *The art of computer programming. Volume* 1: *fundamental algorithms* (Addison Wesley, Reading, MA, 1968).

[3] N.J. Nilsson, *Problem solving methods in artificial intelligence* (McGraw-Hill, New York, 1971).

[4] A.J. Scott, *Combinatorial programming, spatial analysis and planning* (Methuen, London, 1971).

[5] P.D. Stephens, "The IMP language and compiler", *Computer Journal* 17 (1974) 216–233.

Mathematical Programming Study 13 (1980) 58–67.
North-Holland Publishing Company

CONSTRUCTING TIMETABLES FOR SPORT COMPETITIONS

J.A.M. SCHREUDER

Twente University of Technology, Enschede, The Netherlands

Received 1 February 1980

The purpose of this paper is to present an algorithm for constructing feasible solutions of sport competitions e.g. soccer. After the definition of what is meant by a competition, necessary and sufficient conditions for the existence of a competition is proved with the aid of edge-colouring of complete graphs. Feasible timetables can be found by constructing an oriented edge-colouring.

For a fair competition it is necessary to find for each club a Home-and-Away Pattern, such that each club plays as few as possible two or more Home-(or Away)-matches after each other. Based on graph-theoretical results found by de Werra, an algorithm is presented. This algorithm constructs timetables, where no club plays more than once two Home-(or Away)-matches after each other in a half-competition.

Key words: (Canonical) 1-factorization, Chromatic Index, Competition, Complete Graphs, Complete Matchings, Hamiltonian Cycle/Circuit, (Oriented) Edge-colouring, Scheduling, Sport, Timetable.

1. Introduction

In sports a lot of competitions are played between different clubs e.g. in soccer, baseball, hockey etc. A hard problem for the competition-leaders is how to find a good competition schedule such that the wishes of the clubs, public and properties of a "fair" competition can be honoured. Up to now the competition-leaders construct such a schedule by hand, a most time-consuming and frustrating work.

In general such problems are called Time-tabling or scheduling, and a number of publications are known, specially School Time-tabling, see e.g. Aust [1], Brittian and Farley [2] and De Werra [8]. In this paper we shall restrict ourselves to the construction of competition schemes for sporting clubs. A guide for the demands of the competitions will be the Dutch major soccer league. A lot of different demands and heuristic solutions are described by Cain [4] and Campbell and Chen [5].

The general demands of competition are that each club plays a home- and an away-match against all other clubs and that as much as possible, clubs play one match in each competitionweek. The whole competition must take no more weeks than necessary of course, because a lot of other obligations can exist like tournaments, holidays etc.

It is common in most competitions that, when a club plays a home-match in the first half of the competition, he plays the away-match against the same club in the second half of the competition. If we restrict ourselves to construct only the half-competition, we can always find the second half of the competition just by resetting the home-matches in away-matches and otherwise. This is done by Dutch major soccer league.

Of course, there are competitions which cannot be obtained in this way, e.g. if we allow that two clubs play their home-match and away-match in the same half. We shall show at the end of the next paragraph how to change the definitions of the half-competition in dealing with the whole competition.

2. Definition of a half-competition

If we restrict ourselves to an even number of clubs: n, then $(n-1)$ competition-weeks are necessary for playing all the demanded matches in a half-competition. If the number of clubs is odd, we add one fictitious club. The club who is playing against the fictitious club in a certain week, is free during that week.

We define a half-competition as follows:

(a) each club plays a home-match or an away-match against all the other clubs,

(b) each club plays one match (home or away) in each competitionweek, or in each competitionweek all the clubs play a match.

Introducing 0–1 variables

$$x_{ijt} = \begin{cases} 1 & \text{if club } i \text{ plays a home-match against club } j \\ & \text{in competitionweek } t, \\ 0 & \text{else,} \end{cases}$$

we can formally define a half-competition for n clubs (n even) as a zero–one matrix, of which $\frac{1}{2}n(n-1)$ coefficients are 1,

$$X = [x_{ijt}], \quad i = 1, 2, \dots, n; j = 1, 2, \dots, n; t = 1, 2, \dots, n-1$$

satisfying the following conditions.

For each (i, j, t) $i \neq j$: if $x_{ijt} = 1$, then:

(1) $\quad x_{i\beta t} = 0 \quad \forall \beta \neq j, \beta = 1, 2, \dots, n,$

(2) $\quad x_{\alpha i t} = 0 \quad \forall \alpha, \alpha = 1, 2, \dots, n,$

(3) $\quad x_{\alpha j t} = 0 \quad \forall \alpha \neq i,$

(4) $\quad x_{j\beta t} = 0 \quad \forall \beta,$

(5) $\quad x_{ij\gamma} = 0 \quad \forall \gamma \neq t, \gamma = 1, 2, \dots, n-1,$

(6) $\quad x_{ji\gamma} = 0 \quad \forall \gamma,$

(7) $\quad x_{iit} = 0 \quad \forall i, t.$

Interpretation of these conditions is obvious.
Condition (1): club i cannot play more home matches in week t,
Condition (2): club i cannot play an away-match in week t,
Condition (3): club j cannot play more away-matches in week t,
Condition (4): club j cannot play an home-match in week t,
Condition (5): club i cannot play more home-matches against club j in the half-competition,
Condition (6): club j cannot play a home-match against club i in the half-competition,
Condition (7): a club cannot play against himself.

Theorem 1. *A zero–one matrix* $X = [x_{ijt}]$ *(of appropriate dimension) is a half-competition if and only if*

$$(i) \qquad \sum_{i=1}^{n} (x_{ijt} + x_{jit}) = 1 \quad \forall j, t,$$

$$(ii) \qquad \sum_{t=1}^{n-1} (x_{ijt} + x_{jit}) = 1 \quad \forall i \neq j.$$

(A) If a zero–one matrix X satisfies (i) and (ii), then $x_{ijt} = 1$ implies (1), (2), ... , (7).

Proof. We can rewrite (i):

$$\sum_{\alpha \neq i} x_{\alpha jt} + x_{ijt} + \sum_{\beta} x_{j\beta t} = 1, \quad \alpha, \beta = 1, \dots, n.$$

If $x_{ijt} = 1$, then $\sum_{\alpha \neq i} x_{\alpha jt} = \sum_{\beta} x_{j\beta t} = 0 \Rightarrow$(3) and (4).
Another way of rewriting (i):

$$\sum_{\alpha} x_{\alpha it} + \sum_{\beta \neq j} x_{i\beta t} + x_{ijt} = 1, \quad \alpha, \beta = 1, \dots, n$$

If $x_{ijt} = 1$, then $\sum_{\alpha} x_{\alpha it} = \sum_{\beta \neq j} x_{i\beta t} = 0 \Rightarrow$(1) and (2).
Rewrite (ii):

$$\sum_{\gamma \neq t} x_{ij\gamma} + x_{ijt} + \sum_{\gamma} x_{ji\gamma} = 1, \quad \gamma = 1, \dots, (n-1).$$

If $x_{ijt} = 1$, then $\sum_{\gamma \neq t} x_{ij\gamma} = \sum_{\gamma} x_{ji\gamma} = 0 \Rightarrow$(5) and (6).
Rewrite (i):

$$\sum_{\alpha \neq j} x_{\alpha jt} + \sum_{\beta \neq j} x_{j\beta t} + x_{jjt} + x_{jjt} = 1.$$

It is impossible that $x_{jjt} = 1$, therefore $x_{jjt} = 0 \Rightarrow$(7).

(B). If a zero–one matrix X satisfies (1), (2), ... , (7), then $x_{ijt} = 1$ implies (i) and (ii).

Proof. We can write:

$$\sum_{\alpha=1}^{n} x_{\alpha jt} + \sum_{\beta=1}^{n} x_{j\beta t} = \sum_{\alpha \neq i} x_{\alpha jt} + x_{ijt} + \sum_{\beta} x_{j\beta t} \quad \forall j, t.$$

If $x_{ijt} = 1$, then $x_{\alpha jt} = 0 \, \forall \alpha \neq i$, also $\sum_{\alpha \neq i} x_{\alpha jt} = 0$, and $x_{j\beta t} = 0 \, \forall \beta$, also $\sum_{\beta} x_{j\beta t} = 0$.
Therefore: $\sum_{i=1}^{n}(x_{ijt} + x_{jit}) = 1 \, \forall j, t \quad \Rightarrow$ (i).
In the same way we can write:

$$\sum_{\gamma=1}^{n-1}(x_{ij\gamma} + x_{ji\gamma}) = \sum_{\gamma \neq t} x_{ij\gamma} + x_{ijt} + \sum_{\gamma} x_{ji\gamma} \quad \forall i \neq j.$$

If $x_{ijt} = 1$, then $x_{ij\gamma} = 0 \, \forall \gamma \neq t$, also $\sum_{\gamma \neq t} x_{ij\gamma} = 0$, and $x_{ji\gamma} = 0$, also $\sum_{\gamma} x_{ji\gamma} = 0$.
Therefore: $\sum_{t=1}^{n-1}(x_{ijt} + x_{jit}) = 1 \, \forall i \neq j \Rightarrow$ (ii).

For a whole competition consisting of $2(n-1)$ competitionweeks we have only to change (a) such that each club plays one home-match and one out-match against all the other clubs. If we leave out the mathematically definition (6), in (ii) the term x_{jit} and let $t = 1, 2, \ldots, 2(n-1)$, then Theorem 1 is still valid.

3. Existence of a half-competition

A possibility for proving the existence of a half-competition is to make use of graph theoretical results as described by Fiorini and Wilson [6]. We define a *graph* G to be a pair $(V(G), E(G))$, where $V(G)$ is a finite non-empty set of elements called *vertices*, and $E(G)$ is finite set of distinct unordered pairs of distinct elements of $V(G)$ called *edges*. An independent set of edges, or *matching*, in G is a set of edges of G no two of which are adjacent. An independent set of edges which includes every vertex of G is called a *complete matching* in G.

A graph in which every two vertices are adjacent is called a *complete graph*; the complete graph with n vertices will be denoted by K_n.

The *chromatic index* $X'(G)$ of a graph G is the minimum number of colours needed to colour the edges in G in such a way that no two adjacent edges are assigned to the same colour.

An *edge-colouring* of a graph G is a colouring of the edges of G in such a way that all the edges meeting at the same vertex have a different colour. G is said to be *K-edge-colourable* if K is any integer satisfying $X'(G) \leq K$.

Theorem 2. *The chromatic index of* K_n $(n \geq 2)$ *is given by* $X'(K_n) = n-1$, *if n is even.*

Proof. We note first that $X'(K_n) \geq n-1$, since every vertex of K_n has valency $n-1$. We can prove that $X'(K_n) = n-1$ by explicitly constructing an $(n-1)$-colouring of the edges of K_n. If $n = 2$, this is trivial. If $n > 2$, we choose any vertex v, and colour the edges of $K_n - v$ (a complete graph on $n-1$ vertices). To effect such a colouring, we place the vertices of $K_n - v$ in the form of a regular $(n-1)$-gon, and colour the edges around the boundary using a different colour for each edge. The remaining edges can be coloured by assigning to each one the same colour as that used for the boundary edge parallel to it, see Fig. 1(a). At each vertex there will be exactly one colour missing and these missing colours will all be different. The edges of K_n incident to v can be coloured using these missing colours, see Fig. 1(b). $\Rightarrow K_n$ is $(n-1)$-edge-colourable.

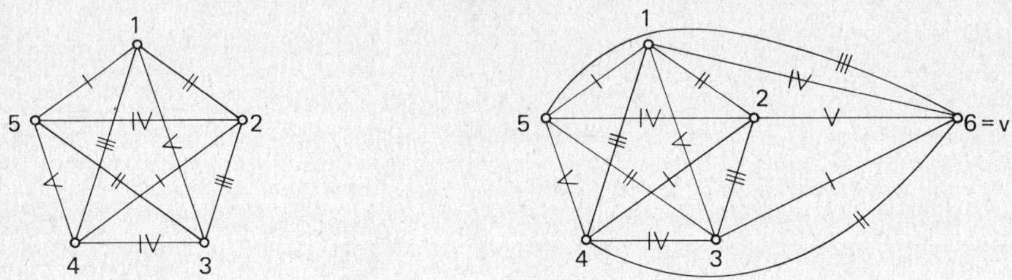

Fig. 1(a). Edge-colouring for $K_n - v$ $(n = 6)$. (b) Edge-colouring for K_n $(n = 6)$.

With each $(n-1)$-edge-colouring C of K_n, there are associated $n-1$ subgraphs $G_1, G_2, \ldots, G_{n-1}$ defined as follows. The set of vertices of G_t is the same as the set of vertices of K_n and the set of edges of G_t is the set of those edges of K_n that have colour t. Each G_t is a matching since the edges of the same colour are not adjacent. In fact, each G_t is a complete matching, see Fig. 2, since if there is a vertex i of zero-valency in G_t then C is not a $(n-1)$-edge-colouring (since vertex i has $n-1$ valency in K_n and there are only $n-2$ colours in C different from t).

We can interpret the results of the graph theory as described above for the competition problem. Each vertex of K_n represents a club (n clubs), index i or j. Each club plays exactly one match against all the other clubs. Therefore, we can represent the possible matches by the edges $\{i, j\}$ of K_n.

A complete matching G_t represents the matches in a competitionweek. All the complete matchings $(n-1)$ of a possible $(n-1)$-edge-colouring of K_n represent a half-competition. Define

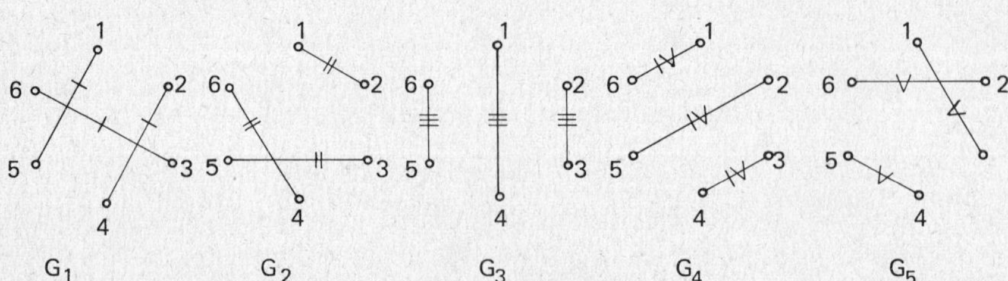

Fig. 2. Five complete matchings with no edge in common if K_6 is coloured with five colours (I, ... , V).

$$x_{ijt} = \begin{cases} 1 & \text{if } [i<j \text{ and edge } \{i,j\} \in G_t], \\ 0 & \text{if } [j<i \text{ and edge } \{i,j\} \in G_t] \text{ or } [\text{edge } \{i,j\} \notin G_t]. \end{cases}$$

Consider (i): For a (j, t) there exists exactly one edge $\{i, j\} \in G_t$. Let $i < j$, then $x_{ijt} = 1$. This implies $x_{\alpha jt} = 0\ \forall \alpha \neq i, \alpha = 1, \dots, n$ and $x_{j\beta t} = 0\ \forall \beta, \beta = 1, \dots, n$. Therefore $\sum_{\alpha \neq i} x_{\alpha jt} = 0$ and $\sum_{\beta} x_{j\beta t} = 0$.

$$\text{(i)} \qquad \sum_{i=1}^{n} (x_{ijt} + x_{jit}) = \sum_{\alpha \neq i} x_{\alpha jt} + x_{ijt} + \sum_{\beta} x_{j\beta t} = 1 \quad \forall j, t.$$

If $j < i$, then $x_{jit} = 1$. For the same reasoning as described before (i) is valid.

Consider (ii): For each $i \neq j$ there exists exactly one edge $\{i, j\}$ in all the complete matchings $G_1, G_2, \dots, G_{n-1}$. With a same reasoning as used for (i), (ii) is valid.

Therefore, a 0–1 matrix $X = [x_{ijt}]$ with conditions (i) and (ii) constructed as an $(n-1)$-edge-colouring of K_n, is a half-competition.

4. Home- and Away Pattern (HAP)

Since a match between two clubs i and j is played either in the home city of club i or j, we can represent their match by an oriented edge $\{i, j\}$. If club i plays at *home*, we represent the match by the arc(j, i) oriented *from j to i* (club j *goes to* club i). We say that arc(j, i) represents a *home-match* for club i and an *away-match* for club j.

A set of $n-1$ disjunct complete matchings (or 1-factorization) of K_n together with an orientation for each edge, gives an *oriented* $(n-1)$*-edge-colouring* of K_n denoted by $\vec{G}_t$.

If we construct a half-competition for six clubs based on the constructive proof of Theorem 2 and its consequences, see e.g. Harary [7], we would get a timetable as in Fig. 3. The meaning of the Figs. 3(a), (b) and (c) in connection with the zero-one matrix $X = [x_{ijt}]$ is the following:

Fig. 3(a): $x_{ijt} = 1 \Leftrightarrow$ in $\vec{G}_t$: $\overleftarrow{ij}$ (in week t club i plays at home against club j).

Fig. 3(b): $x_{ijt} = 1 \Rightarrow$ in week t: $i = 1$ (home-match) and $j = 0$ (away-match).

Fig. 3(c): $x_{ijt} = 1 \Leftrightarrow S$: $s_{ij} = t$ same as Fig. 3(a).

(a)

$\vec{G}_1$: $\overleftarrow{1\ 5}$ $\overleftarrow{2\ 4}$ $\overleftarrow{3\ 6}$

$\vec{G}_2$: $\overleftarrow{1\ 2}$ $\overleftarrow{3\ 5}$ $\overleftarrow{4\ 6}$

$\vec{G}_3$: $\overleftarrow{1\ 4}$ $\overleftarrow{2\ 3}$ $\overleftarrow{5\ 6}$

$\vec{G}_4$: $\overleftarrow{1\ 6}$ $\overleftarrow{2\ 5}$ $\overleftarrow{3\ 4}$

$\vec{G}_5$: $\overleftarrow{1\ 3}$ $\overleftarrow{2\ 6}$ $\overleftarrow{4\ 5}$

(b)

	Week				
Club	1	2	3	4	5
1	1	$\underline{1}$	$\underline{1}$	$\underline{1}$	$\underline{1}$
2	1	0	$\underline{1}$	$\underline{1}$	$\underline{1}$
3	1	$\underline{1}$	0	1	0
4	0	1	0	$\underline{0}$	1
5	0	$\underline{0}$	1	0	$\underline{0}$
6	0	$\underline{0}$	$\underline{0}$	$\underline{0}$	$\underline{0}$

(c)

		Away				
Club	1	2	3	4	5	6
1	×	2	5	3	1	4
H 2		×	3	1	4	5
o 3			×	4	2	1
m 4				×	5	2
e 5					×	3
6						×

Fig. 3. Timetable for six clubs. (a) Oriented 5-edge-colouring (G) e.g. in week 1 club 1 plays at home against club 5. (b) Home-and-Away Pattern (HAP) e.g. clubs 1, 2, 3 play at home in week 1. (c) Schedule (S) e.g. club 1 plays at home against club 2 in week 2.

In a competition one tries to construct as much as possible alternating home- and away-matches for each club. As can be seen from Fig. 3(b), that schedule is far from ideal. We define a *break*, if a club plays two times after each other a home- (or away-)match. In the HAP of Fig. 3(b) we see that there are 14 breaks (underlined).

Using graph-theoretical results, De Werra [9] was able to prove that it is possible to construct timetables for half-competitions with exactly n-2 breaks (n clubs, n even). The breaks are ordered in such a way that 2 clubs play an ideal alternating HAP and the remaining $n-2$ clubs play only once two home- or away-matches after each other.

The general idea of finding a HAP with $n-2$ breaks is combining 2 complete matchings of an $(n-1)$-edge-colouring in such a way, that they form a Hamiltonian circuit of length n. Because each graph K_n (n even) is the edge-sum of $(\frac{1}{2}n-1)$ disjunct Hamiltonian cycles and a 1-factor, see [3], proper orientation of the edges gives a $(n-1)$-oriented-edge-colouring with $n-2$ breaks.

For example see Fig. 2; if we orient the edges in the order 1-5-3-6-4-2-1, G_1 and G_2 form a Hamiltonian circuit, the same counts for G_3 and G_4.

More formally we want to find a *canonically feasible* 1-*factorization* of K_n. A 1-factorization $(G_1, G_2, \dots, G_{n-1})$ of K_n will be called canonical, if for $j = 1, 2, \dots, n-1$ G_j is defined by

$$G_j = \{[n, j]\} \cup \{[j+k, j-k];\ k = 1, 2, \dots, n-1\}$$

where the numbers $j+k, j-k$ are expressed as one of the numbers $1, 2, \dots, n-1 \bmod (n-1)$. A 1-factorization is canonically feasible if each arc is oriented in order to obtain an oriented colouring $(\vec{G}_1, \vec{G}_2, \dots, \vec{G}_{n-1})$ of K_n. The consequence is that in each consecutive two columns of the HAP-matrix, there are at least two breaks except in the first two columns.

Based on this results of De Werra [9] the following algorithm gives a HAP with exactly $n-2$ breaks.

Algorithm

Take a club from the n clubs, say n.
Form the set I_{n-1} of the remaining $n-1$ clubs:

$$I_{n-1} = \{1, 2, \dots, n-1\}$$

Step 0: TEST 1 := 'TRUE'; $t := i := 0$; (t: weeknumber; $i \in I_{n-1}$}

$$x_{ijt} := 0\ \forall i, j, t$$

Step 1: $t := t+1$
Step 2: $i := i+1$

IF $i = \frac{1}{2}n$ THEN TEST 1 := NOT TEST 1
IF TEST 1 = 'TRUE'
THEN $x_{n,i,t}$:= 1; TEST 1 := 'FALSE'
ELSE $x_{i,n,t}$:= 1; TEST 1 := 'TRUE'
TEST 2 := 'TRUE'

Step 3: $\alpha := \beta := i$
DO $(\frac{1}{2}n - 1)$ times
$\alpha := \alpha - 1$ modulo $(n-1)$
$\beta := \beta + 1$ modulo $(n-1)$
IF TEST 2 = 'TRUE'
THEN $x_{\alpha,\beta,t}$:= 1; TEST 2 := 'FALSE'
ELSE $x_{\beta,\alpha t}$:= 1; TEST 2 := 'TRUE'

Step 4: If $t < n-1$ GOTO Step 1
STOP

If we use the algorithm for constructing a timetable for six clubs, we would get Fig. 4. In Fig. 4(b) week 1 and 2 and week 4 and 5 form Hamiltonian circuits.

$\vec{G}_1$: $\overleftarrow{6\ 1}$ $\overrightarrow{2\ 5}$ $\overleftarrow{3\ 4}$
$\vec{G}_2$: $\overrightarrow{6\ 2}$ $\overrightarrow{3\ 1}$ $\overleftarrow{4\ 5}$
$\vec{G}_3$: $\overrightarrow{6\ 3}$ $\overrightarrow{4\ 2}$ $\overleftarrow{5\ 1}$
$\vec{G}_4$: $\overleftarrow{6\ 4}$ $\overrightarrow{5\ 3}$ $\overleftarrow{1\ 2}$
$\vec{G}_5$: $\overrightarrow{6\ 5}$ $\overrightarrow{1\ 4}$ $\overleftarrow{2\ 3}$

(a)

Club	Week 1	2	3	4	5	6	7
1	0	1	0	1	0	*1*	*0*
2	0	1	$\underline{1}$	0	1	$\underline{\mathit{1}}$	*0*
3	1	0	1	$\underline{1}$	0	$\underline{\mathit{0}}$	*1*
4	0	1	0	$\underline{0}$	1	$\underline{\mathit{1}}$	*0*
5	1	0	1	0	1	*0*	*1*
6	1	0	$\underline{0}$	1	0	$\underline{\mathit{0}}$	*1*

(b)

Club	Week 1	2	3	4	5	6
1	×	4	2	*10*	*8*	*6*
2	*9*	×	5	3	*6*	2
3	*7*	*10*	×	1	4	3
4	5	*8*	6	×	2	*9*
5	3	1	*9*	7	×	5
6	1	*7*	*8*	4	*10*	×

(c)

Fig. 4. Timetable for six clubs. (a) Oriented 5-edge-colouring (G). (b) HAP. (c) Schedule for whole competition; $t = 1,2,\dots,2n-2$.

If we look at Fig. 4(b) (HAP), we see that there are now only 4 breaks. We construct a whole competition as in Fig. 4(c) by resetting the home- and away-matches for each club. The total minimum number of breaks is $3n-6$ (here: 12). We don't allow a break in the second and $(n-1)$th week, because otherwise there would be two consecutive breaks in the whole competition. If we don't want 4 home- or away-matches in 5 weeks (e.g. $0\ \underline{0}\ 1\ 0\ \underline{0}$) in a whole competition, there should also be no breaks in the fourth and $(n-3)$th week (for 10 or more clubs). All this has been taken care of by the second line in Step 2 of the algorithm.

5. Conclusion

An interesting consequence of constructing a half-competition by a canonically feasible 1-factorization is that for an odd number of clubs there are no breaks; strik e.g. in Fig 4(b) club 6 out. If we not only reset the home- and away-matches for the second half of the competition, but also mirror the matches (e.g. week 1 $6 \leftarrow 1$, week 10 $6 \rightarrow 1$ or week 5 $6 \leftarrow 5$, week 6 $6 \rightarrow 5$), we could get a competition with no breaks.

In the present situation the Dutch competition leader starts with an HAP (Home- and Away-Pattern) with $3n-6$ breaks. Then he has to assign the real clubs to the club-numbers *and* decide in which week each two pair of clubs play their match. With our algorithm he has only to assign the clubs to the numbers. Of course he has to take into account a number of requirements. One requirement is that some clubs are situated so close together, that they don't want to play their home-matches in the same week. For each two pair of clubs this is easy to achieve. The reason is that the HAP of each club is always complementary to that of one other club (property of oriented-edge-colouring, De Werra [9]); see e.g. Fig. 4(b): club 1 and 5, 2 and 6 and 3 and 4 have complementary patterns. Another requirement could be that in some weeks some clubs cannot play at home. Furthermore we have to construct an interesting competition schedule such that in consecutive weeks each club gets opponents of different strength.

Acknowledgment

I wish to thank Dr. M. Vlach for his substantial comments on the subject.

References

[1] R.J. Aust, "An improvement algorithm for school timetabling", *Computer Journal* 19 (1979) 339–343.

[2] J.N.G. Brittain and F.J.N. Farley, "College timetable construction by computer", *Computer Journal* 14 (1971) 361–365.

[3] M. Behzad, G. Chartrand and L. Lesniak-Foster, *Graphs and digraphs* (Prindle, Weber & Smidt International Series, Boston, IL) p. 168.

[4] W.O. Cain Jr., "The computer-assisted heuristic approach used to schedule the major league

baseball clubs", in: S.P. Ladany and R.E. Machol, eds., *Optimal strategies in sports* (North-Holland, Amsterdam, 1977) pp. 32–41.
[5] R.T. Campbell and D.S. Chen, "A minimum distance basketball scheduling problem", in: R.E. Machol, S.P. Ladany and D.G. Morrison, eds., *Management science in sports* (North-Holland, New York, 1976) pp. 15–25.
[6] S. Fiorini and R.J. Wilson, "Edge-Colourings of graphs", *Research Notes in Mathematics* 16 (Pitman Publishing, London, 1977).
[7] F. Harry, *Graph theory* (Addison-Wesley, Reading, MA, 1977).
[8] D. De Werra, "Constructing of school timetables by flow methods", *Infor Journal* 9 (1971) 12–22.
[9] D. De Werra, "Scheduling in sports", O.R. Working Paper 45, Département de Mathématiques, Ecole Polytechnique Federale de Lausanne, Switzerland (1979).

Mathematical Programming Study 13 (1980) 68–77.
North-Holland Publishing Company

THE RECONSTRUCTION OF LATIN SQUARES WITH APPLICATIONS TO SCHOOL TIMETABLING AND TO EXPERIMENTAL DESIGN

A.J.W. HILTON

Department of Mathematics, University of Reading, Whiteknights, Reading, Great Britain

Received 1 February 1980

When trying to construct a school timetable, a good first step might seem to be to construct an outline timetable in which all History teachers are counted together, all French teachers are counted together, etc., all classes of each year group are counted together, and in which the preliminary division is into days rather than lessons. Having constructed an outline timetable satisfying one's main outline requirements, one might then go on to develop this outline timetable into a complete timetable.

This paper shows that, for the appropriate kind of timetable, this is always a feasible approach. The mathematics is cast in terms of a theorem on reconstructing latin squares. This theorem also shows that frequency squares, or F-squares, used in Statistics may always be obtained by identifying various symbols in a latin square.

Key words: Edge-colouring, Frequency Squares, Graphs, Latin Squares, Outline Rectangle, Reduction, Timetable.

1. Introduction

When trying to draw up a school timetable so many quirks and difficulties may be imposed that the task becomes impossible. In order to lessen these the task is usually performed by someone with high authority such as Mr. Blenkinshaw, the Deputy Headmaster, and this has the obvious effect of reducing some absolute requirements to merely desirable features of a timetable. Mr. Blenkinshaw's task is then to manufacture a timetable incorporating as many desirable features as possible.

Mr. Blenkinshaw may find that a good approach is to start off by lumping all the History teachers together, all the French teachers, etc., to count all classes of each year group together, and to divide the week into days rather than lessons. He may, alternatively, find that some variation on this theme is more effective. Having constructed a suitable outline timetable he may then go on to develop it into a proper timetable. Whether or not this approach is useful to him depends very much on the kind of features of the timetable he finds most important or desirable. Some difficulties that he may encounter, and some ways of getting round them, are suggested in the last section.

To turn to the mathematical theorem we wish to discuss. A *latin square* L of side n is an $n \times n$ matrix on symbols $1, \ldots, n$ in which each row and each column contain each symbol exactly once. From L one may derive a matrix B by amalgamating various of the rows, various of the columns and various of the symbols in a way described in detail in the next section. Since it was derived from a latin square, B will satisfy various numerical constraints. We show that, conversely, to each matrix B satisfying these numerical constraints there is a latin square from which it could have been derived.

A latin square is not unlike a school timetable. The number of masters and the number of classes are not very different, and for the purposes of this discussion we may suppose that a few dummy classes are introduced so that these numbers are the same. Normally there are many classes which are not taught by any given master, and it is this feature which makes F-squares, or frequency squares, studied in Statistics a more natural model of a timetable. An *F-square* is an $n \times n$ matrix filled with at most n symbols, in which each symbol occurs precisely the same number of times in each row and each column. Thus one could reserve a given symbol, say ϕ, to stand for the fact that a master and a class do not meet. If all masters teach the same number of classes and all classes are taught by the same number of masters, then an F-square might be a good model of a timetable.

2. Forming outline rectangles from latin squares

A composition A of a positive integer n is a sequence $(a_1, \dots, a_r)$ of positive integers such that $a_1 + \cdots + a_r = n$. Let $P = (p_1, \dots, p_u)$, $Q = (q_1, \dots, q_v)$ and $S = (s_1, \dots, s_w)$ be three compositions of n. The *reduction modulo* (P, Q, S) of a latin square L of side n on the symbols $1, \dots, n$ is obtained from L by amalgamating rows $p_1 + \cdots + p_{i-1} + 1, \dots, p_1 + \cdots + p_i$, columns $q_1 + \cdots + q_{j-1} + 1, \dots, q_1 + \cdots + q_j$ and symbols $s_1 + \cdots + s_{k-1} + 1, \dots, s_1 + \cdots + s_k$ for $1 \le i \le u$, $1 \le j \le v$ and $1 \le k \le w$.

More precisely, for $1 \le \xi \le n$, $1 \le \lambda \le u$ and $1 \le \mu \le v$, let $x(\lambda, \mu, \xi)$ be the number of times that symbol ξ occurs in the set of cells $\{(i, j): p_1 + \cdots + p_{\lambda-1} + 1 \le i \le p_1 + \cdots + p_\lambda,\ q_1 + \cdots + q_{\mu-1} + 1 \le j \le q_1 + \cdots + q_\mu\}$ and, for $1 \le k \le w$, let

$$x_k(\lambda, \mu) = x(\lambda, \mu, r_1 + \cdots + r_{k-1} + 1) + \cdots + x(\lambda, \mu, r_1 + \cdots + r_k).$$

Then the reduction modulo (P, Q, S) or L is a $u \times v$ matrix B whose cells are filled from the symbols $\tau_1, \dots, \tau_w$ (say) and in which cell (λ, μ) contains $\tau_k\, x_k(\lambda, \mu)$ times.

Given a reduction modulo (P, Q, S) of a latin square L, clearly this itself may be further reduced by further amalgamation of rows, columns and symbols.

We now illustrate this reduction process with an example: Let the given latin square L be as shown in Diagram 1.Then $n = 12$ (to make this accord with the description above we may let π be 10, ∞ be 11 and e be 12). Let u, v and w be 3, 4 and 4 respectively and let $(p_1, p_2, p_3) = (4, 4, 4)$, $(q_1, q_2, q_3, q_4) = (5, 3, 3, 1)$ and

Diagram 1

e	1	2	3	4	5	6	7	8	9	π	∞
∞	π	e	9	1	7	2	8	3	6	4	5
π	e	∞	1	9	2	8	3	4	7	5	6
9	∞	π	8	e	1	7	2	5	3	6	4
5	6	7	π	8	e	4	1	9	2	∞	3
1	2	3	4	5	6	9	∞	7	8	e	π
4	9	1	5	2	π	3	6	∞	e	7	8
2	3	4	e	6	8	5	π	1	∞	9	7
6	7	8	∞	π	3	e	5	2	4	1	9
7	8	5	6	3	4	∞	9	π	1	2	e
8	4	9	2	7	∞	π	e	6	5	3	1
3	5	6	7	∞	9	1	4	e	π	8	2

$(s_1, s_2, s_3, s_4) = (4, 3, 3, 2)$. Let I be the composition consisting of a sequence of the appropriate length of 1's.

The reduction of L modulo (P, Q, I) is given in Diagram 2. In this diagram, there is not intended to be any significance in the way the symbols are arranged in each cell.

Diagram 2

<table>
<tr><td>1 1 1 2 3
4 8 9 9 9
π π π ∞ ∞
∞ e e e e</td><td>1 2 2
2 3 5
6 7 7
7 8 8</td><td>3 3 4
4 5 5
6 6 7
8 9 π</td><td>4
5
6
π</td></tr>
<tr><td>1 1 2 2 2
3 3 4 4 4
5 5 5 6 6
7 8 9 π e</td><td>1 3 4
5 6 6
8 9 π
π ∞ e</td><td>1 2 7
7 8 9
9 ∞ ∞
∞ e e</td><td>3
7
8
π</td></tr>
<tr><td>2 3 3 4 5
5 6 6 6 7
7 7 7 8 8
8 9 π ∞ ∞</td><td>1 3 4
4 5 9
9 π ∞
∞ e e</td><td>1 1 2
2 3 4
5 6 8
π π e</td><td>1
2
9
e</td></tr>
</table>

Finally we reduce this modulo (I, I, S); this means that we replace 1, 2, 3 and 4 by $a = \tau_1$; 5, 6 and 7 by $\beta = \tau_2$; 8, 9 and π by $\delta = \tau_3$; and ∞ and e by $\epsilon = \tau_4$. We obtain Diagram 3. This is then the reduction modulo (P, Q, S) of L.

Diagram 3

α	α	α	α	α	α	α	α	α	α	α	α
α	δ	δ	δ	δ	α	α	β	α	β	β	β
δ	δ	δ	ε	ε	β	β	β	β	β	β	β
ε	ε	ε	ε	ε	β	δ	δ	δ	δ	δ	δ
α	α	α	α	α	α	α	α	α	α	β	α
α	α	α	α	α	β	β	β	β	δ	δ	β
β	β	β	β	β	δ	δ	δ	δ	ε	ε	δ
β	δ	δ	δ	ε	δ	ε	ε	ε	ε	ε	ε
α	α	α	α	β	α	α	α	α	α	α	α
β	β	β	β	β	α	β	δ	α	α	α	α
β	β	β	δ	δ	δ	δ	ε	β	β	δ	δ
δ	δ	δ	ε	ε	ε	ε	ε	δ	δ	ε	ε

We now define an outline rectangle. Let C be a $u \times v$ matrix filled with the w symbols $\tau_1, \dots, \tau_w$ in which each cell may be occupied by more than one symbol and in which each symbol may occur more than once. For $1 \le \lambda \le u$, $1 \le \mu \le v$ and $1 \le \nu \le w$, let ρ_λ be the number including repetitions of symbols which occur in row λ, let c_μ be the number including repetitions of symbols which occur in column μ and let σ_ν be the number of times τ_ν appears in C. Then C is called an *outline rectangle* if for some integer n the following properties are obeyed for each λ, μ, ν such that $1 \le \lambda \le u$, $1 \le \mu \le v$ and $1 \le \nu \le w$:

(i) n divides each ρ_λ, c_μ and σ_ν;

(ii) cell (λ, μ) contains $(1/n^2)\, \rho_\lambda c_\mu$ symbols (including repetitions);

(iii) the number of times τ_ν appears in row λ is $(1/n^2)\, \rho_\lambda \sigma_\nu$;

(iv) the number of times τ_ν appears in column μ is $(1/n^2)\, c_\mu \sigma_\nu$.

We remark that F-squares are the special case of the outline rectangles considered here in which $u = v = \rho_1 = \cdots = \rho_\mu = c_1 = \cdots = c_v = n$. Thus our main theorem, Theorem 3, is a theorem about F-squares.

Theorem 1. *The reduction modulo (P, Q, S) of L is an outline rectangle and has the further properties*

(v) $$(p_1, \dots, p_u) = \left(\frac{\rho_1}{n}, \dots, \frac{\rho_u}{n}\right);$$

(vi) $$(q_1, \dots, q_v) = \left(\frac{c_1}{n}, \dots, \frac{c_v}{n}\right);$$

(vii) $$(s_1, \dots, s_w) = \left(\frac{\sigma_1}{n}, \dots, \frac{\sigma_w}{n}\right);$$

(viii) $\sum_{\lambda=1}^{u} \rho_\lambda = \sum_{\mu=1}^{v} c_\mu = \sum_{\nu=1}^{w} \sigma_\nu = n^2.$

Proof. Obvious.

3. Forming latin squares from outline rectangles

In this section we show that any outline rectangle could have been formed from some latin square by reduction modulo (P, Q, S) for some suitable compositions P, Q and S. Our main tool is a theorem of de Werra which we explain and prove in the next sub-section.

3.1. De Werra's theorem

The graph theory terminology we employ here is standard if it is used without explanation and may be found in [7] or [9].

Let G be a graph with vertex set V and edge set E. Let G contain multiple edges but no loops. An *edge-colouring* of G with colours $1, \ldots, k$ is a partition of E into k mutually disjoint subsets $C_1, \ldots, C_k$. Thus $C_i \cap C_j = \phi$ $(1 \le i < j \le k)$ and $C_1 \cup \cdots \cup C_k = E$. An edge has colour i if it belongs to C_i. Note that we do not make the usual requirement that two edges having the same colour do not have a vertex in common.

Given an edge-colouring of G, for each $v \in V$ let $C_i(v)$ be the set of edges on v of colour i, and, for each $u, v \in V$, $u \neq v$, let $C_i(u, v)$ be the set of edges joining u and v of colour i.

An edge-colouring of G is called *equitable* if, for all $v \in V$,

(a) $$\max_{1 \le i < j \le k} \Big| |C_i(v)| - |C_j(v)| \Big| \le 1,$$

and is called *balanced* if, in addition, for all $u, v \in V$, $u \neq v$,

(b) $$\max_{1 \le i < j \le k} \Big| |C_i(u, v)| - |C_j(u, v)| \Big| \le 1.$$

Thus an edge-colouring is balanced if the colours occur as uniformly as possible at each vertex and if the colours are shared out as uniformly as possible on each multiple edge. D. de Werra [4, 5, 6] proved the following important theorem. We give here for completeness, and because it is short, a proof due to Andersen [2].

Theorem 2 (de Werra). *For each $k \ge 1$, any finite bipartite graph has a balanced edge-colouring with k colours.*

Proof. Colour the edges of the graph in such a way that (b) is satisfied; condition (b) only affects each multiple edge by itself, so this is clearly possible. We then modify the colouring to make (a) be satisfied without violating (b). Suppose that at some vertex v,

$$\max_{1\le i<j\le k} \left| |C_i(v)| - |C_j(v)| \right| > 1.$$

We may suppose that this maximum is attained for colours 1 and 2 and that $|C_1(v)| > |C_2(v)| + 1$. Let P be a maximal chain $v = v_0, e_1, v_1, e_2, v_2, \dots, e_h, v_h$ (where e_i is an edge joining v_{i-1} to v_i and $e_i \ne e_j$ if $i \ne j$) such that

(I) e_1 has colour 1,

(II) $e_1, \dots, e_h$ are coloured alternately 1 and 2,

(III) $|C_1(v_i, v_{i+1})| = |C_2(v_i, v_{i+1})| + 1$ if i is even, and $|C_2(v_i, v_{i+1})| = |C_1(v_i, v_{i+1})| + 1$ if i is odd,

(IV) P uses only one edge from each multiple edge.

(Note that the same vertex may occur several times in P.)

Then $h \ne 0$ because v has some neighbour v_1 for which $|C_1(v, v_1)| = |C_2(v, v_1)| + 1$, since $|C_1(v)| > |C_2(v)| + 1$. Also $v_h \ne v_0$, because if $v_j = v_0$, then j is even as the graph is bipartite, so when j edges have been traversed, both colours have occurred the same number of times in total on the multiple edges incident with v_0 used so far and so the chain can be continued since $|C_1(v)| > \|C_0(v)| + 1$.

Interchanging the two colours 1 and 2 on the chain P clearly does not violate (b), it reduces the number of pairs of colours for which $\left| |C_i(v)| - |C_j(v)| \right|$ was maximal by at least one and it does not affect

$$\max_{1\le i<j\le k} \left| |C_i(v_t)| - |C_j(v_t)| \right|$$

if $0 < t < h$.

If h is odd, then e_h has colour 1, so the maximality of P implies that the number of multiple edges (v_h, x) on v_h for which

$$|C_1(v_h, x)| = |C_2(v_h, x)| + 1$$

exceeds the number of multiple edges on v_h for which

$$|C_2(v_h, x)| = |C_1(v_h, x)| + 1$$

by at least one, so the colour 1 occurs at least once more than the colour 2 at v_h. Thus

$$\max_{1\le i<j\le k} \left| |C_i(v_h)| - |C_j(v_h)| \right|$$

will not be increased. A similar statement is true if h is even.

Repeated application of the argument then proves Theorem 2.

3.2. *Application of de Werra's theorem to forming latin squares from outline rectangles*

We now state and prove the main theorem.

Theorem 3. *To each outline rectangle C there is a latin square L and compositions P, Q and S such that C is the reduction of L modulo (P, Q, S).*

Proof. First observe that if x denotes the number of entries in C, then

$$x=\sum_{\lambda=1}^{u}\rho_\lambda=\sum_{\mu=1}^{v}c_\mu=\sum_{\lambda=1}^{u}\sum_{\mu=1}^{v}\frac{1}{n^2}\rho_\lambda c_\mu.$$

Therefore

$$x=\frac{1}{n^2}\sum_{\lambda=1}^{u}\rho_\lambda\left(\sum_{\mu=1}^{v}c_\mu\right)=\frac{x}{n^2}\sum_{\lambda=1}^{u}\rho_\lambda=\frac{x^2}{n^2},$$

so $x=n^2$.

We next observe that the outline rectangle can be represented as a family of triplets (x, y, z) where each occurrence of each symbol in each cell of C corresponds to exactly one triplet, the first coordinate denoting the row the cell lies in, the second the column and the third the number of the symbol itself. Thus if cell (λ, μ) contains symbol τ_ν we obtain the triple (λ, μ, ν). There are therefore n^2 triples, counting repetitions. The conditions (ii), (iii) and (iv) now take on the more symmetrical form:

(ii)′ (λ, μ) occurs as the first pair in $(1/n^2)\ \rho_\lambda c_\mu$ triples;
(iii)′ (λ, ν) occurs as the first and last entries in $(1/n^2)\ \rho_\lambda\sigma_\nu$ triples;
(iv)′ (μ, ν) occurs as the last pair in $(1/n^2)\ \lambda_\mu\sigma_\nu$ triples.

Because of this symmetry we may without loss of generality confine the explanation to the case when $u<n$ and show that C can be obtained from a $(u+1)\times v$ outline rectangle C' by amalgamating the cells of two rows (so that any pair of cells in these two rows which are in the same column are identified) or, in other words, by reduction modulo (P^*, I, I), where P^* is a composition with one term 2, the rest all ones. Repeated application of this argument first on the rows, then on the columns and finally on the symbols will show that C can be obtained from an $n\times n$ outline rectangle on n symbols, i.e. a latin square, by reduction modulo (P, Q, S) for some compositions P, Q and S.

Since $u\neq n$, n divides $\rho_1, \dots, p_\mu$ and $\sum_{\lambda=1}^{u}\rho_\lambda/n=n$ there is at least one λ for which $\rho_\lambda/n>1$. We may assume without loss of generality that $\rho_u/n\geq 2$. We wish to form an outline rectangle C' by splitting the last row of C into two new rows. We construct a bipartite graph C with vertex classes $\{\gamma_1, \dots, \gamma_v\}$ and $\{\tau_1, \dots, \tau_w\}$ where the vertex γ_μ is joined to the vertex τ_ν by y edges if and only if the symbol τ_ν occurs y times in cell (u, μ) of C. Then the degree of γ_μ is the number, including repetitions, of symbols in the cell, namely $(1/n^2)\,\rho_u c_\mu$, and the degree of the vertex τ_ν is the number of times τ_ν occurs in row u of C, namely $(1/n^2)\,\rho_u\sigma_\nu$.

We now give G an equitable edge-colouring with ρ_u/n colours. Let C_1 be the set of those edges coloured with colour 1. Then each γ_μ has exactly

$$\frac{n}{\rho_u}\left(\frac{1}{n^2}\rho_u c_\mu\right)=\frac{1}{n}c_\mu$$

edges of colour 1 on it and each vertex τ_ν has exactly

$$\frac{n}{\rho_u}\left(\frac{1}{n^2}\rho_u\sigma_\nu\right)$$

edges of colour 1 on it. Now split row u of C into two rows u' (to be row $u+1$ of C') and u'' (to be row u of C') by placing a symbol τ_ν in cell (u',μ) x times if and only if there are x edges of colour 1 joining the vertices γ_μ and τ_ν, and by placing τ_ν in cell (u'',μ) y times if and only if there are y edges of colours different from 1 joining the vertices γ_μ and τ_ν.

We now check that C' is an outline rectangle. Let $\rho'_\lambda=\rho_\lambda$ $(1\le\lambda<u)$, $\rho'_u=\rho_u-n$ and $\rho'_{u+1}=n$. Let $c'_\mu=c_\mu$ $(1\le\mu\le v)$ and $\tau'_\nu=\tau_\nu$ $(1\le\nu\le n)$. Then clearly n divides each of ρ'_λ, c'_μ and σ'_ν. Cells (u,μ) and $(u+1,\mu)$ of C' contain, respectively,

$$\frac{1}{n^2}c_\mu(\rho_u-n)=\frac{1}{n^2}c'_\mu\rho'_u$$

and

$$\frac{1}{n}c_\mu=\frac{1}{n^2}\rho'_{u+1}c'_\mu$$

symbols, including repetitions. Each symbol τ_ν occurs

$$\frac{1}{n^2}\sigma_\nu(\rho_u-n)=\frac{1}{n^2}\rho'_u\sigma'_\nu$$

times in row u of C' and

$$\frac{\sigma_\nu}{n}=\frac{1}{n^2}\rho'_{u+1}\sigma'_\nu$$

times in row $u+1$. Thus in these cells and rows conditions (ii)–(iv) applied to C' are satisfied and also they are clearly satisfied in all columns and all other cells and rows.

This proves Theorem 3.

Perhaps we should point out that in the proof above we only used the fact that the edge-colouring of G was equitable and that we did not need to use the idea of balance. Balance would have been needed for a more general theorem which would imply for example that the matrix

1 1 1	1 2 2
1 2 2	1 1 1

could have been derived by reducing the numbers modulo 2 from a matrix of the type where there is a bound on the number of repetitions in any cell such as

1 3 3	3 2 2
3 2 2	1 3 3

rather than from one such as

3 3 3	1 2 2
1 2 2	3 3 3

.

For some results in this direction see [1] and [2].

4. Concluding remarks

The study of F-squares in Statistics has been particularly concerned with the existence or non-existence of orthogonal sets of F-squares (see [8]). We feel that our Theorem 3 should have a bearing in the future on how this study is conducted. Thus it might be that instead of investigating the existence of an orthogonal mate to a given F-square, one might possibly adopt some measure of near-orthogonality, and look for nearly orthogonal mates to the latin squares from which the given F-square could have been derived.

So far as school timetabling is concerned what we would like to emphasize is that the general approach of forming outline timetables and then developing them into complete ones is sound. There are a number of difficulties, and perhaps we might mention briefly possible ways of overcoming them.

If a given master has to meet a given class x times during the week, one might look for a generalization of an F-square in which the cell at the intersection of the column representing the master and the row representing a class is a cell with x symbols in it (see Fig. 1).

One could then prepare a preliminary chart giving the number of symbols to be placed in each cell. From that one could find an outline timetable with the appropriate number of entries everywhere, and finally this could be developed into a complete timetable.

Another way of dealing with this type of problem occurs if many masters have to meet many classes for the same number x of times in the week. If these x meetings can be arranged for the same x hours for each of the masters and each of the classes, then these x meetings could be counted as a unit for timetable purposes.

Some mathematical analogues of Theorem 3 have been investigated in [1] (see also [2]); in particular the analogue when $P = Q$ and $S = I$. It seems quite likely that other analogues await discovery.

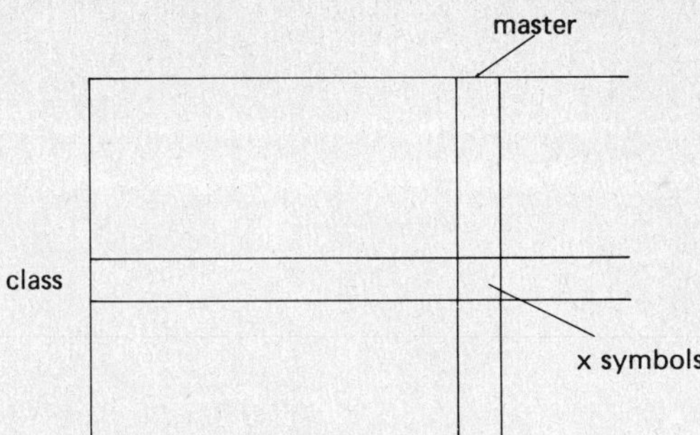

Fig. 1.

References

[1] L.D. Andersen and A.J.W. Hilton, "On constructing and embedding generalized latin rectangles", *Discrete Mathematics*, to appear.
[2] L.D. Andersen and A.J.W. Hilton, "Generalized latin rectangles", in: R.J. Wilson, ed., *Proceedings of the one-day conference on combinatorics at the Open University* (Pitman, London, 1979) pp. 1–17.
[3] J. Denes and A.D. Keedwell, *Latin squares and their applications* (Academic Press, New York, London, 1974).
[4] D. de Werra, "Balanced schedules", *Infor Journal* 9 (1971) 230–237.
[5] D. de Werra, "A few remarks on chromatic scheduling", in; B. Roy, ed., *Combinatorial programming: methods and applications* (D. Reidel, Dordrecht, Holland, 1975) pp. 337–342.
[6] D. de Werra, "On a particular conference scheduling problem", *Infor Journal* 13 (1975) 308–315.
[7] F. Harary, *Graph theory* (Addison-Wesley, Reading, MA, 1969).
[8] A. Hedayat and E. Seiden, "F-square and orthogonal F-square design: a generalization of latin squares and orthogonal latin square design", *Annals of Mathematical Statistics* 41 (1970) 2035–2044.
[9] R.J. Wilson, *Introduction to graph theory* (Oliver and Boyd, Edinburgh, 1972).

Mathematical Programming Study 13 (1980) 78–87.
North-Holland Publishing Company

AN ALGORITHM FOR THE SINGLE MACHINE SEQUENCING PROBLEM WITH PRECEDENCE CONSTRAINTS

C.N. POTTS

Department of Mathematics, University of Keele, Keele, Staffordshire, Great Britain

Received 1 February 1980

The single machine sequencing problem is considered, in which each job has a processing time and a weight, and there are precedence constraints on the jobs. The objective is to find a sequence of jobs which minimises the weighted sum of completion times. A new lower bound is derived and used in a branch and bound algorithm. Computational results for up to forty jobs are given.

Key words: Completion Time, Dominance Rules, Graph, Heuristic, Precedence Constraints, Processing Time, Search Strategy, (Single Machine) Sequencing, Zero–One Programming.

1. Introduction

The problem may be stated as follows. Each of n jobs is to be processed without interruption on a single machine. At any instant the machine can process only one job and there is to be no idle time between processing jobs. Precedence constraints on the jobs are represented by a directed acyclic graph G. The vertices of G represent the jobs and if a directed path from vertex i to vertex j exists, then job i must be processed before job j.

Each job i has a positive processing time p_i and a weight w_i. Given any sequence of jobs, the completion time C_i for any job i can be determined assuming that processing starts at time zero. The objective is to find a sequence of jobs which minimises the weighted sum of completion times $\sum_{i=1}^{n} w_iC_i$. Henceforth, the weighted sum of completion times is denoted by WSCT.

Smith [9] showed that the problem can be solved in O(n log n) steps, when there are no precedence constraints, by sequencing the jobs in non-increasing order of w_i/p_i. Following various generalizations by Conway et al. [1], Horn [2] and Sidney [8] to include various precedence constraint structures, Lawler [4] derived an algorithm for the problem with series parallel precedence constraints. This also requires O(n log n) steps provided that the decomposition tree of G is given. For general precedence constraints, Lawler [4] and Lenstra and Rinnooy Kan [5] have shown that the problem is NP-hard. A branch and bound algorithm, using a lower bound obtained from the solution of a linear assignment problem, has been proposed by Rinnooy Kan et al. [7] for problems with a more general objective function. Finally Morton and Dharan [6] compared the computational results produced by three heuristics. Two of these heuristics consistently generated solutions close to the optimum.

Some terms that are used in later sections are now introduced. The *transitive closure* of the directed graph G is the graph obtained by adding all arcs (i, j) to G whenever there is a directed path from vertex i to vertex j. (If the arc (i, j) already exists in G, then no new arc (i, j) is added.) The *transitive reduction* of G is the graph obtained by deleting all arcs (i, j) from G whenever there is a directed path from vertex i to vertex j which does not include the arc (i, j) itself. The *inverse* of G is the graph obtained by reversing the directions of all arcs. The *adjacency matrix* of the transitive closure of G is the $n \times n$ matrix $A = (a_{ij})$, where $a_{ij} = 1$ if there is an arc (i, j) in the transitive closure of G and $a_{ij} = 0$ otherwise. If the arc (i, j) exists in the transitive closure of G, then i is a *predecessor* of j and j is a *successor* of i. If the arc (i, j) exists in the transitive reduction of G, then i is a *direct predecessor* of j and j is a *direct successor* of i.

In this paper we propose a branch and bound algorithm. Section 2 contains the branching rule and a lower bounding rule is derived in Section 3. The complete algorithm is given in Section 4 including details of the implementation of these rules. Computational experience is presented in Section 5 which is followed by some concluding remarks in Section 6.

2. Branching rule

In this section a branching rule is given which partitions the set of feasible solutions to the original problem into subsets. Some dominance theorems are given which eliminate some of these subsets. We then show that the subproblems generated by this branching rule have the same characteristics as the original problem.

Our branching rule is similar to that used by Kurisu [3] in his branch and search algorithm for minimising the maximum completion time in the two-machine flow-shop problem with precedence constraints. Essentially, at each branching a job is selected and is sequenced either first, last, immediately after another given job or immediately before another given job.

We now state some results that have appeared in the literature which act as dominance rules to reduce the number of branches of the search tree. The corollaries relate to the corresponding results for the equivalent inverse problem in which the objective is to maximise WSCT subject to the precedence constraints defined by the inverse graph.

Theorem 1 (Horn [2]). *If job i has no predecessors and $w_i/p_i \geq w_j/p_j$ for all jobs j, then an optimum sequence exists in which job i is sequenced first.*

Corollary 1. *If job i' has no successors and $w_{i'}/p_{i'} \leq w_j/p_j$ for all jobs j, then an optimum sequence exists in which job i' is sequenced last.*

Theorem 2 (Morton and Dharan [6]). *If job i has at least one predecessor and $w_i/p_i \geq w_j/p_j$ for all jobs j, then there exists an optimum sequence in which job i is sequenced immediately after one of its direct predecessors.*

Corollary 2. *If job i' has at least one successor and $w_{i'}/p_{i'} \leq w_j/p_j$ for all jobs j, then there exists an optimum sequence in which job i' is sequenced immediately before one of its direct successors.*

The details of our branching rule are now given. Firstly job i and job i' are found such that $w_i/p_i \geq w_j/p_j$ and $w_{i'}/p_{i'} \leq w_j/p_j$ for all jobs j. If i or i' is not uniquely defined, then an arbitrary choice is made. Let k and k' denote the numbers of direct predecessors and direct successors of i and i' respectively. Then there are four cases to be considered.

(a) If $k = 0$, then job i can be sequenced first (Theorem 1).

(b) If $0 = k' < k$, then job i' can be sequenced last (Corollary 1).

(c) If $0 < k \leq k'$, then k branches of the search tree are formed. In each branch job i is sequenced immediately after one of its direct predecessors (Theorem 2).

(d) If $0 < k' < k$, then k' branches of the search tree are formed. In each branch job i' is sequenced immediately before one of its direct successors (Corollary 2).

When these rules are applied, as few branches as possible are added to the search tree at each branching. Cases (a) and (b) check whether a single branch can be added by sequencing a job either first or last according to Theorem 1 or Corollary 1. If not, we either apply Theorem 2 in case (c) or Corollary 2 in case (d), depending on which adds the smaller number of branches to the search tree.

Morton and Dharan follow this procedure in their tree-optimal heuristic except that only one branch of the search tree is considered in cases (c) and (d). In (c) a branch is chosen corresponding to a direct predecessor j of i with the smallest value of w_j/p_j, while in (d) a branch is chosen corresponding to a direct successor j' of i' with the largest value of $w_{j'}/p_{j'}$. This tree-optimal heuristic has the advantage that if no branches are ignored in (c) or (d), then an optimum solution is necessarily generated.

Finally in this section we show that, following a branching, new problems are produced with the same characteristics as the original but with one less job. This means that our branching rule can be applied at any node of the search tree.

If job i is sequenced first or job i' is sequenced last, then its contribution to WSCT can be evaluated after which it can be removed from the list of jobs and its vertex deleted from G. If job j and job i are to be sequenced consecutively, then these jobs can be replaced by a single composite job (j, i) with processing time $p_j + p_i$ and weight $w_j + w_i$. This will increase WSCT by $p_i w_j$ but otherwise leaves the problem unchanged [2]. The precedence graph for the new problem is obtained by performing the following condensation of G.

(a) Vertex j and vertex i are deleted and a single new vertex (j, i) is added.

(b) For each arc (k, j) or (k, i) in G, where $k \neq j$, an arc $(k, (j, i))$ is added.

(c) For each arc (j, k) or (i, k) in G, where $k \neq i$, an arc $((j, i), k)$ is added.

The case in which job i' and j' are to be sequenced consecutively is treated similarly.

3. Lower bounds

For simplicity a lower bound is derived for the initial problem before branching occurs. The generalizations needed for calculating bounds at lower levels of the search tree are straightforward.

The problem is first formulated as a zero–one programming problem. We define, for any jobs i and j, the zero–one variable

$$x_{ij} = \begin{cases} 0 & \text{if job } i \text{ is sequenced after job } j, \\ 1 & \text{otherwise.} \end{cases}$$

It follows from this definition that $x_{ii} = 1$ for all jobs i. The values of some x_{ij} are implied by the precedence constraints (if $a_{ij} = 1$ in the adjacency matrix of the transitive closure of G, then $x_{ij} = 1$ and $x_{ji} = 0$), while others need to be determined. Now the completion of processing job j occurs at time $\sum_{i=1}^{n} p_i x_{ij}$. Thus the problem may be written

$$\text{minimise} \sum_{j=1}^{n} \sum_{i=1}^{n} p_i w_j x_{ij}, \tag{1}$$

$$\begin{aligned} \text{subject to} \qquad x_{ij} &\geq a_{ij}, \quad i,\ j = 1, \ldots, n, && (2)\\ x_{ij} + x_{ji} &= 1, \quad i,\ j = 1, \ldots, n,\ i \neq j, && (3)\\ x_{ij} + x_{jk} + x_{ki} &\leq 2, \quad i,\ j, k = 1, \ldots, n,\ i \neq j,\ j \neq k,\ k \neq i, && (4)\\ x_{ii} &= 1, \quad i = 1, \ldots, n,\\ x_{ij} &= 0 \text{ or } 1, \quad i,\ j = 1, \ldots, n. \end{aligned}$$

The constraints (2) ensure that $x_{ij} = 1$ whenever the precedence constraints imply that job i is to be sequenced before job j. The relationship that any job i is to be sequenced either before or after any other job j is represented by (3). If the leading diagonal elements of the matrix $X = (x_{ij})$ are ignored, then X may be regarded as the adjacency matrix of a directed graph G_X. The constraints (4) ensure that G_X contains no cycles. When all constraints are satisfied, the transitive reduction of G_X is a single chain defining a sequence of jobs.

Clearly, if the constraints (4) are relaxed, then a lower bound on WSCT is obtained. The resulting problem can be expressed in terms of the variables above the leading diagonal in X using the constraints (3). The objective becomes

$$\text{minimise} \sum_{i=1}^{n-1} \sum_{j=i+1}^{n} (p_i w_j - p_j w_i) x_{ij},$$

apart from an additive constant term which has been omitted. Other than the zero–one restrictions, there are no constraints on those variables x_{ij} whose values are not fixed by the precedence constraints. Thus the relaxed problem is solved by assigning values to these variables as follows:

$$x_{ij} = \begin{cases} 0 & \text{if } p_i w_j - p_j w_i \geq 0, \\ 1 & \text{otherwise.} \end{cases} \tag{5}$$

The corresponding value of the objective function (1) is denoted by B. In the absence of precedence constraints, the relationship with Smith's rule is apparent. However, the following problem with 3 jobs illustrates that the values of our variables do not necessarily define a sequence. In this first example $p_1 = 3$, $p_2 = 2$, $p_3 = 1$, $w_1 = 2$, $w_2 = 1$, $w_3 = 1$ and the graph G, shown in Fig. 1, specifies that job 2 is to be sequenced before job 3. Therefore $x_{23} = 1$ and applying (5) we obtain $x_{12} = 1$ and $x_{13} = 0$, which yields $B = 16$. The corresponding graph G_X is also shown in Fig. 1. Clearly the arcs (1, 2), (2, 3) and (3, 1) form a cycle in G_X. The notation (123) is used to denote such a cycle henceforth. We next propose a method of improving the lower bound B by eliminating some of the cycles in G_X.

Suppose that $D = \{D_1, \ldots, D_d\}$, where D_k $(k = 1, \ldots, d)$ is the set of arcs forming some cycle of G_X. We define $\bar{D}_k$ $(k = 1, \ldots, d)$ as the largest subset of D_k containing no arcs of the transitive closure of G. We also define

$$c_k = \min_{(i,j) \in \bar{D}_k} \{p_j w_i - p_i w_j\}, \quad k = 1, \ldots, d. \tag{6}$$

From (5), each c_k is non-negative. Our improved lower bound is now stated.

Theorem 3. *If $\bar{D}_1, \ldots, \bar{D}_d$ are disjoint, then*

$$B_D = B + \sum_{k=1}^{d} c_k$$

is a lower bound for WSCT.

Proof. If G_X is to define a sequence it must contain no cycles. Therefore the

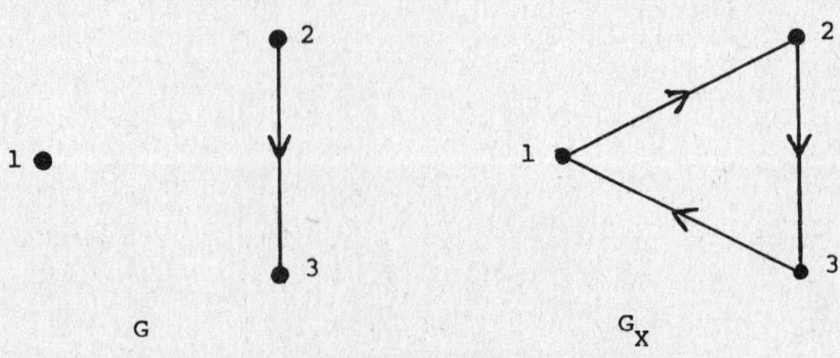

Fig. 1. G and G_X for the first example.

direction of at least one of the arcs in each of $D_1, \ldots, D_d$ must be reversed. As the precedence constraints must be satisfied, the minimum additional contributions to WSCT are $c_1, \ldots, c_d$. This completes the proof.

It should be noted that if $\bar{D}_1, \ldots, \bar{D}_d$ are not disjoint, then reversing the direction of one arc may eliminate more than one of the d cycles, in which case B_D may not be a lower bound for WSCT. This may be illustrated by the following problem with 4 jobs. In this second example $p_1 = 8$, $p_2 = 1$, $p_3 = 2$, $p_4 = 4$, $w_1 = 1$, $w_2 = 1$, $w_3 = 1$, $w_4 = 1$ and the graph G, shown in Fig. 2, specifies that job 1 is to be sequenced before job 2 and job 3. Therefore $x_{12} = 1$ and $x_{13} = 1$ and applying (5) we obtain $x_{14} = 0$, $x_{23} = 1$, $x_{24} = 1$, $x_{34} = 1$ which yields $B = 39$. The graph G_X is shown in Fig. 2 and has cycles (124) and (134). If $D_1 = \{(1,2), (2,4), (4,1)\}$ and $D_2 = \{(1,3), (3,4), (4,1)\}$, then $\bar{D}_1 = \{(2,4), (4,1)\}$ and $\bar{D}_2 = \{(3,4), (4,1)\}$. The arc (4, 1) is common to $\bar{D}_1$ and $\bar{D}_2$. Now applying (6) we obtain $c_1 = 3$ and $c_2 = 2$ which yields $B_D = 44$. Thus B_D is greater than 43, the minimum value of WSCT given by the sequence (1, 2, 3, 4).

The details of the method for finding an appropriate set D are discussed now. Rather than attempt to find a set D which gives the highest value of B_D, we present a rule which generates a good bound without excessive computational requirements. Consequently the search for D is restricted to cycles containing only three arcs. Firstly an arc (i, j) of the transitive closure of G is chosen arbitrarily and a vertex k is found such that (ijk) is a cycle in G_X and such that the increase in B_D by including the arcs of this cycle in D is as large as possible. Then the arcs (i, j), (j, k) and (k, i) are deleted from G_X. This procedure is repeated until either $B_D \geq U$, where U is an upper bound on WSCT, or no further increase in B_D is possible.

4. The algorithm

It is well-known that computation can be reduced by using a heuristic method to find a good solution to act as an upper bound on WSCT prior to the

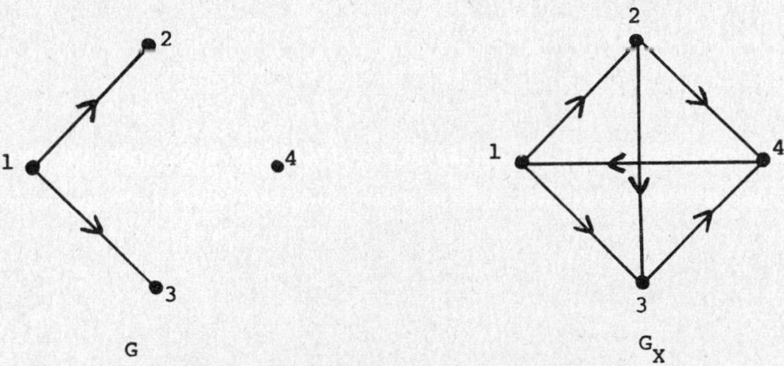

Fig. 2. G and G_X for the second example.

application of a branch and bound algorithm. In our algorithm the tree-optimal heuristic of Morton and Dharan, outlined in Section 2, is used.

Suppose that a parent node has been chosen from which to branch. Then the branching rule described in Section 2 is applied to generate the descendants. If this rule sequences one job either first or last, then the lower bound for the descendant is the same as that of the parent and need not be recalculated. In other cases, for each descendant, the transitive closure of the precedence graph is required for the computation of the lower bound B. At the root node the computation of the transitive closure requires $O(n^3)$ steps. Elsewhere, the transitive closure of G at the parent node can be used in finding the transitive closure at the descendants as follows. If the branching rule specifies that job j and job i are to be sequenced consecutively, then the transitive closure is updated by adding the arc (h, j) whenever an arc (h, i) exists $(h \neq j)$ and by adding the arc (h, k) whenever the arcs (h, i) and (j, k) exist $(h \neq j,\ k \neq i)$. Then vertex i is deleted and vertex j now corresponds to the composite job (j, i). The lower bound B for this descendant can now be simply computed. If there are s unscheduled jobs at the parent node, then the calculation of B, including the updating procedure described above, requires $O(s^2)$ steps for each descendant. However, to compute B_D requires a further $O(s^3)$ steps for each descendant.

Finally our search strategy is given. A newest active node search is used which selects a node from which to branch which has the smallest lower bound amongst nodes in the most recently created subset.

5. Computational experience

The two algorithms, using the lower bounds B and B_D respectively, were tested on problems with 20, 30 and 40 jobs. For each job i, an integer processing time p_i from the uniform distribution [1, 100] and an integer weight w_i from the uniform distribution [1, 10] were generated. In the precedence graph G, each arc (i, j) with $i < j$ was included with a given probability P. For each selected value of n, twenty problems were generated for each of the P values 0.05, 0.1, 0.15, 0.2, 0.3, 0.5 and 0.75.

The algorithms were coded in FORTRAN IV and run on a CDC 7600 computer. Computational results are given in Table 1. Whenever a problem was not solved after 10 000 nodes had been generated, computation was abandoned for that problem. Thus in some cases the figures given in Table 1 are lower bounds on average computation times and average numbers of nodes.

For problems with twenty or thirty jobs the lower bound B gives smaller computation times than B_D even though the search trees have on average over twice as many nodes for some values of P. On this evidence it would appear that the extra computation needed to calculate B_D is not profitable. However, for problems with forty jobs and with the P values 0.15, 0.2 and 0.3, B_D yields

Table 1
Computational results

		Lower bound B		Lower bound B_D	
n	P	Average computation time[a]	Average number of nodes	Average computation time[a]	Average number of nodes
20	0.05	0.011	6.2	0.013	5.2
	0.1	0.016	18.1	0.023	15.5
	0.15	0.027	40.4	0.036	23.3
	0.2	0.030	48.3	0.045	27.4
	0.3	0.046	67.9	0.064	34.3
	0.5	0.025	22.7	0.034	12.8
	0.75	0.015	3.1	0.016	0.7
30	0.05	0.039	29.0	0.058	26.6
	0.1	0.247	276.6	0.316	150.1
	0.15	0.412	417.8	0.520	197.7
	0.2	0.418	413.0	0.528	171.7
	0.3	0.311	284.1	0.515	149.2
	0.5	0.080	43.2	0.134	23.1
	0.75	0.041	1.9	0.045	0.9
40	0.05	1.220	716.8	1.510	404.5
	0.1	4.453[b]	2369.2[b]	5.486	1157.1
	0.15	6.012[b]	2814.6[b]	4.478	793.6
	0.2	7.900[c]	3774.6[c]	7.635	1383.5
	0.3	3.798	1573.6	3.023	422.2
	0.5	0.461	193.3	0.549	54.2
	0.75	0.100	10.6	0.138	5.7

[a] Times are in CPU seconds.
[b] Lower bounds because of one unsolved problem.
[c] Lower bounds because of four unsolved problems.

smaller average computation times than B. This reduction in computation by using B_D is greater than the figures indicate for $P = 0.15$ and $P = 0.2$ due to the unsolved problems.

It appears that our algorithms solve problems with up to thirty jobs efficiently. Unfortunately some of the search trees become large when there are forty jobs. For a given number of jobs, the problems with small and large values of P are easiest. This is expected because in the limiting case $P = 0$ the jobs can be sequenced using Smith's rule, while there is only one feasible sequence for $P = 1$. It seems likely that the hard problems in which the search trees are largest occur when the transitive reduction of the precedence graph contains the most arcs.

Some experiments were performed to compare the assignment bound of Rinnooy Kan et al. with our lower bound B. The assignment gave a value that was smaller or equal to B for almost every problem and required a longer computation time. On this evidence our lower bound is more efficient.

6. Concluding remarks

Both of our algorithms are satisfactory for solving small and medium sized problems. The branching rule successfully limits the size of the search tree in such cases. A disadvantage of the proposed algorithm is that the lower bound is not exact for the case of series parallel precedence constraints. It is possible that computation could be reduced by checking at each node whether the precedence constraints are series parallel. If at any node they are series parallel, Lawler's algorithm can be applied, after which this node can be discarded.

Another obvious attempt at improving efficiency is to reduce the computational requirement of B_D by choosing the set D differently. Although this may yield a smaller lower bound, average computation times may be reduced. Also worthy of investigation is the lower bound obtained by applying Lawler's algorithm to the problem in which precedence constraints are defined by a series parallel subgraph of the original precedence graph. However, the best method of finding this subgraph is unknown.

Acknowledgment

The author is grateful to the Mathematisch Centrum, Amsterdam and The Royal Society for helping to finance a visit to the Mathematisch Centrum where some of this research was undertaken. Many useful discussions with B.J. Lageweg, J.K. Lenstra and A.H.G. Rinnooy Kan are gratefully acknowledged.

References

[1] R.W. Conway, W.L. Maxwell and L.W. Miller, *Theory of scheduling* (Addison-Wesley, Reading, MA, 1967).

[2] W.A. Horn, "Single-machine job sequencing with treelike precedence ordering and linear delay penalties", *Society for Industrial and Applied Mathematics Journal of Applied Mathematics* 23 (1972) 189–202.

[3] T. Kurisu, "Two-machine scheduling under arbitrary precedence constraints", *Journal of the Operations Research Society of Japan* 20 (1977) 113–131.

[4] E.L. Lawler, "Sequencing jobs to minimise total weighted completion time subject to precedence constraints", *Annals of Discrete Mathematics* 2 (1978) 75–90.

[5] J.K. Lenstra and A.H.G. Rinnooy Kan, "Complexity of scheduling under precedence constraints", *Operations Research* 26 (1978) 22–35.

[6] T.E. Morton and B.G. Dharan, "Algoristics for single-machine sequencing with precedence constraints", *Management Science* 24 (1978) 1011–1020.

[7] A.H.G. Rinnooy Kan, B.J. Lageweg and J.K. Lenstra, "Minimising total costs in one-machine scheduling", *Operations Research* 23 (1975) 908–927.

[8] J.B. Sidney, "Decomposition algorithms for single-machine sequencing with precedence relations and deferral costs", *Operations Research* 23 (1975) 283–298.

[9] W.E. Smith, "Various optimizers for single-stage production", *Naval Research Logistics Quarterly* 3 (1956) 59–66.

Mathematical Programming Study 13 (1980) 88–101.
North-Holland Publishing Company

FINDING k EDGE-DISJOINT SPANNING TREES OF MINIMUM TOTAL WEIGHT IN A NETWORK: AN APPLICATION OF MATROID THEORY

Jens CLAUSEN and Lone Aalekjær HANSEN

DIKU, University of Copenhagen, Copenhagen, Denmark

Received 1 February 1980

The by now classical Held and Karp procedure for the travelling salesman problem (TSP) and the "$\frac{3}{2}$"-heuristic of Christofides for the Euclidian TSP are both based on the existence of good algorithms for the minimum spanning tree problem.

The problem of finding k edge-disjoint Hamiltonian circuits of minimum total weight in a network, $k \geq 2$, (by J. Krarup called the peripatetic salesman problem (PSP)), is related to problems of both practical and theoretical importance (reallocation of governmental institutions in Sweden, vulnerability in networks). Trying to generalize the Held and Karp procedure and the "$\frac{3}{2}$"-heuristic to solve the PSP, the problem of finding k edge-disjoint spanning trees of minimum total weight in a network (k-MSTP) arises. This problem can be formulated as finding a minimum weight base in a matroid and hence the greedy algorithm can be applied if appropriate independence testing routines are available.

In this paper, we first introduce the necessary concepts and notation from matroid theory including the sum of matroids, and giving a non-standard proof we establish that the sum of k matroids is a matroid.

By means of the sum of matroids, the k-MSTP is formulated as a matroid problem, and two independence testing routines (both variants of the matroid partition algorithm of J. Edmonds) for the matroid in question are described. These are compared w.r.t. computational complexity and computational behaviour, in the latter case with special emphasis on k-MSTP for large sparse graphs.

Finally, the difficulties arising when applying the above sketched exact and heuristic methods to the PSP are discussed.

Key words: Circuits, Edge-disjoint, Graph, Greedy Algorithm, Hamiltonian Circuit, Heuristic, Matroid (Partition), Minimum Spanning Tree, (Peripatetic) Travelling Salesman.

1. Introduction

The development in computer technology over the past half decade has been tremendous, and one of the keywords in this is computer networks. The classical problems of communication networks are of course found in the field of computer networks, too, including the problem of constructing reliable networks either from scratch or by extending existing networks.

Via the concepts of connectivity and vulnerability graph theory offers some tools applicable for solving the reliability problem. A recent survey of results in these fields is by Christofides and Whitlock [3], in which (among many others) the problem of finding k edge-disjoint Hamiltonian circuits of minimum total weight in a network, $k \geq 2$, is discussed. This was baptised as the Peripatetic Salesman problem (PSP) by Krarup [11], who came across the problem in connection with reallocation of governmental institutions in Sweden. However, in neither of the papers an optimal algorithm for the problem is proposed, and the only heuristic mentioned is for the given graph repeatedly to find the minimum weight Hamiltonian circuit and discard this from the graph until k edge-disjoint Hamiltonian circuits have been found. This heuristic is very poor since the removal of one Hamiltonian circuit from a graph may cause the nonexistence of another, even if the graph is the union of two edge-disjoint Hamiltonian circuits, cf. Example 1.

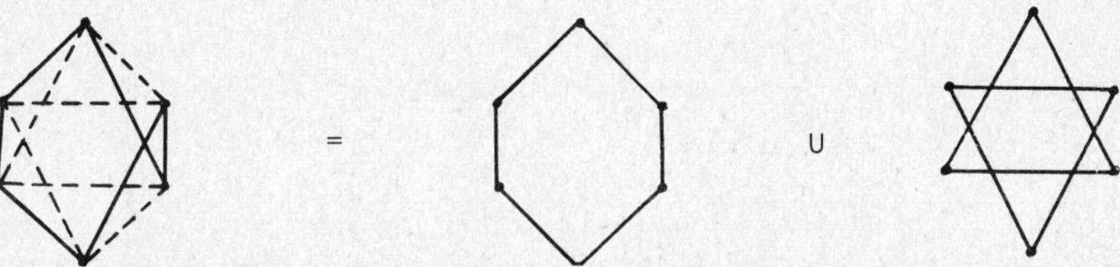

Example 1. A graph, which is both the union of two edge-disjoint Hamiltonian circuits and the union of a Hamiltonian circuit and two triangles.

When k takes on the value 1, the PSP reduces to the well-known Travelling Salesman problem (TSP). Since this problem is NP- complete it is very unlikely that a "good" (i.e. polynomially bounded) algorithm for its solution exists. Nevertheless, reasonably efficient algorithms for the TSP have been devised [8, 9], and for TSP's satisfying the so-called triangle inequality, Christofides [2] has constructed a heuristic, which produces a solution of value not more than $\frac{3}{2}$ times the optimal value in "polynomial" time. Both of these procedures are based on the existence of a good algorithm for determining a spanning tree of minimum weight in a network.

A first step towards generalization of the ideas of Held and Karp, and Christofides is therefore to solve the problem of finding k edge-disjoint spanning trees of minimum total weight in a network (in the following called k-MSTP). In [6] Edmonds noted that in terms of matroid theory, this problem is already solved. The problem can be stated as finding a minimum weight base of a certain matroid, and hence the greedy algorithm applies provided that an independence testing subroutine for this matroid is available.

In this paper we first introduce the necessary concepts and notation from matroid theory including the sum or union of matroids. The k-MSTP is then formulated as the problem of finding a minimum weight base in the sum of k identical graphic matroids, and two independence testing subroutines (both variants of the matroid partition algorithm of Edmonds) for a sum matroid are described. These are then specialized to the matroid in question.

Two versions of the greedy algorithm for solving the k-MSTP have been implemented. Some implementation details w.r.t. data structures are discussed, and the computational behaviour of the algorithms when applied to large sparse graphs is compared.

Finally, the difficulties arising when generalizing the Held and Karp-procedure and the "$\frac{3}{2}$"-heuristic as proposed are discussed.

2. Basic concepts

Let E be a finite nonempty set and $\mathcal{I}$ a collection of subsets of E. The system $(E, \mathcal{I})$ is called an independence system and $\mathcal{I}$ the collection of independent sets iff M satisfies

$$\text{if } I \in \mathcal{I} \text{ and } J \subseteq I, \text{ then } J \in \mathcal{I}. \tag{I1}$$

M is called a matroid iff M in addition satisfies

$$\text{if } I, J \in \mathcal{I} \text{ with } |I| = |J| + 1, \text{ then there exists } r \in I \setminus J \text{ such that } J \cup \{r\} \in \mathcal{I}. \tag{I2}$$

The set E is called the ground set of M.

In the recent years, matroid theory and related areas have attracted much interest from workers in the field of combinatorial optimization due to the applicability of certain parts of the theory (cf. e.g. [12] or [4]). In this section we present the concepts and results to be used throughout; a thorough treatment of matroid theory in general can be found in [17].

Let $M = (E, \mathcal{I})$ be a matroid. A subset of E, which is not independent is called dependent, and a minimal dependent set (i.e. a dependent set, for which any proper subset is independent) is called a circuit of M. If I is an independent set and e an element of $E \setminus I$ it can be shown that in case $I \cup \{e\}$ is dependent, it contains exactly one circuit. This is called the fundamental circuit of I w.r.t. e and is denoted $C(I, e)$.

Consider now a subset A of E. By (I2) it is easily shown that all maximal independent subset of A (i.e. subsets, which are not properly contained in any independent subset of A) have the same number of elements. This is called the rank of A and is denoted $\mathrm{rk}(A)$. The rank of M is defined as $\mathrm{rk}(E)$ and a maximal independent subset of E is called a base of M. The span or closure of A, $\mathrm{sp}(A)$, is the union of A and all elements depending on A, i.e.

$$\mathrm{sp}(A) = A \cup \{e \in E \setminus A \mid \mathrm{rk}(A \cup \{e\}) = \mathrm{rk}(A)\}.$$

A lot of practical problems can be formulated as the problem of finding a maximum weight independent set for an independence system $(E, \mathcal{I})$, in which non-negative weights have been assigned to the elements of E, e.g. the maximum weight spanning tree problem and the linear assignment problem. The simplest approach for solving such a problem is the so-called greedy algorithm. This builds up stepwise the solution by adding at each step to the existing partial solution the element of largest weight not yet considered iff an independent set (i.e. a new partial solution) results. The algorithm is specified below:

Step 1: $S := \emptyset$; $R := E$;
Step 2: Choose e in R of largest weight (ties are resolved arbitrarily);
if $S \cup \{e\} \in \mathscr{I}$ then $S := S \cup \{e\}$;
$R := R \setminus \{e\}$;
if $R = \emptyset$ then STOP else go to Step 2.

It is well-known that the greedy algorithm produces a maximum weight independent set in E regardless of the actual weights iff the independence system considered is a matroid. The complexity of the algorithm is $0(|E| \cdot \log(|E|) + |E| \cdot f(|E|))$ where $f(|E|)$ is the number of computational steps required to test independence in the matroid. Hence the algorithm is good modulo independence testing, i.e. polynomially bounded iff $f(n)$ is bounded above by a polynomial.

It should be noted that the greedy algorithm is equally well suited for finding a minimum weight base of a matroid; "largest" must then be substituted by "smallest". This is due to the fact that all bases are of equal cardinality. Hence it causes no problems that we, though interested in minimization problems, in the following presentation retain tradition and discuss maximization.

3. The sum of matroids

Consider two matroids, $M_1 = (E_1, \mathscr{I}_1)$ and $M_2 = (E_2, \mathscr{I}_2)$, and define the family $\mathscr{I}$ of subsets of $E = E_1 \cup E_2$ by

$$\mathscr{I} = \{I_1 \cup I_2 \mid I_1 \in \mathscr{I}_1, I_2 \in \mathscr{I}_2\}.$$

The system $M = (E, \mathscr{I})$ is called the sum of M_1 and M_2 and is denoted $M_1 \vee M_2$.

Theorem 3.1. *The sum of two matroids M_1 and M_2 as defined above is a matroid.*

Proof. The theorem can be proved in various ways; here we give a simple direct proof due to Mirsky [14] and independently Clausen and Hφholdt [5], which utilizes only the basic properties (I1) and (I2) for matroids.

Obviously, the family $\mathscr{I}$ satisfies (I1). To establish the validity of (I2) for $\mathscr{I}$, let $X, Y \in \mathscr{I}$ with $|X| = |Y| + 1$. Find $X_1 \in \mathscr{I}_1$, $X_2 \in \mathscr{I}_2$ such that $X_1 \cup X_2 = X$ and $X_1 \cap X_2 = \emptyset$. Among the pairs of sets (Y_1, Y_2) which satisfy

$$(*) \qquad Y_1 \in \mathscr{I}_1,\ Y_2 \in \mathscr{I}_2, \quad Y_1 \cup Y_2 = Y, \quad Y_1 \cap Y_2 = \emptyset$$

choose one, for which $|X_2 \cap Y_1| + |X_1 \cap Y_2|$ is minimal (see Fig. 1). Now since $|X| = |Y| + 1$ either $|X_1| > |Y_1|$ or $|X_2| > |Y_2|$. We can assume that $|X_1| > |Y_1|$. Then

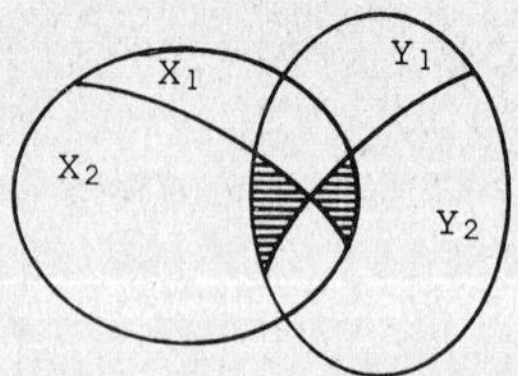

Fig. 1. The shaded area is $(X_2 \cap Y_1) \cup (X_1 \cap Y_2)$.

by (I2) there exists $x \in X_1 \setminus Y_1$ such that $Y_1 \cup \{x\} \in \mathcal{I}_1$. If $x \in Y_2$, then $((Y_1 \cup \{x\}), (Y_2 \setminus \{x\}))$ is a pair of sets satisfying (*) and

$$|X_2 \cap (Y_1 \cup \{x\})| + |X_1 \cap (Y_2 \setminus \{x\})| = |X_2 \cap Y_1| + |X_1 \cap Y_2| - 1$$

contradicting the minimality of $|X_2 \cap Y_1| + |X_1 \cap Y_2|$. Therefore $x \notin Y_2$ and thus $x \in X \setminus Y$ and $\{x\} \cup Y = (Y_1 \cup \{x\}) \cup \{Y_2\} \in \mathcal{I}$, which establishes (I2) for $\mathcal{I}$.

The sum-operation is easily generalized from two to k matroids:

$$M = M_1 \vee \cdots \vee M_k = (E, \mathcal{I})$$

where

$$E = E_1 \cup \cdots \cup E_k,$$
$$\mathcal{I} = \{I_1 \cup \cdots \cup I_k \mid I_1 \in \mathcal{I}_1, \ldots, I_k \in \mathcal{I}_k\}.$$

Note that without loss of generality, we may assume that $M_1, \ldots, M_k$ has a common ground set E.

The partition problem for k matroids $M_1, \ldots, M_k$ with a common ground set E is the following: Let $A \subseteq E$. Is it possible to partition A into k sets $I_1, \ldots, I_k$ s.t. $I_j \subseteq \mathcal{I}_j$ for $j = 1, \ldots, k$? (Edmonds [6]). An equivalent formulation in terms of the sum of matroids is: Let $A \subseteq E$. Is A independent in $M_1 \vee \cdots \vee M_k$?

Edmonds has devised an algorithm for this problem ([6] and [12]) of complexity $0(k \cdot |E|^2 \cdot f(|E|))$, where $f(|E|)$ is the maximum number of computational steps required to test independence in any of $M_1, \ldots, M_k$. This has later been modified by Knuth [10] in order to obtain partitions of the given set with prescribed cardinalities of $I_1, \ldots, I_k$. Both of these algorithms build up stepwise a partition of the set A by determining from the existing partial partition (initially $I_1 = \cdots = I_k = \emptyset$) an element of A, which together with the already "partitioned" elements form a partitionable subset of A (and in Knuth's version satisfies some additional condition). In case A is not partitionable, both of the algorithms produce a subset A' of A for which

$$|A'| > \sum_{i=1}^{k} \mathrm{rk}_i(A').$$

However, the condition

$$\forall A' \subseteq A: \quad |A'| \leq \sum_{i=1}^{k} \mathrm{rk}_i(A')$$

is easily seen to be necessary (and by the previous discussion also sufficient) for A to be partitionable.

4. The maximum weight partitionable subset problem

Let $w: E \to R_+ \cup \{0\}$ be a weight function on the elements of E, and consider now the problem of finding a maximum weight partitionable subset of E w.r.t. the given k matroids $M_1, \dots, M_k$. Since A is partitionable iff A is independent in $M = M_1 \vee \cdots \vee M_k$, this is the problem of determining a maximum weight independent subset in M. Hence the greedy algorithm can be applied, provided a subroutine for independence testing in M is available. Edmonds' algorithm is straightforwardly applicable for this task, and the necessary modifications in Knuth's method turns this into the algorithm of Edmonds.

Assume now that $I \subseteq E$ is independent in M and $e \in E \setminus I$. If $I \cup \{e\}$ is not independent in M, it is in certain applications (electrical network theory) necessary to determine $C(I, e)$, the fundamental circuit in M determined by I and e (see [15]). As mentioned, Edmonds' algorithm produces in this case a subset A of $I \cup \{e\}$, which is not partitionable and which therefore contains $C(I, e)$. However, in most cases, A contains some additional elements. To overcome this difficulty, Petersen [16] has designed yet another version of Edmonds' algorithm, which works "in the opposite direction" of the algorithms of Edmonds and Knuth. Given a partitionable subset I of E and $e \in E \setminus I$, Petersen's algorithm starts searching from e to determine whether $I \cup \{e\}$ is independent in M. If this is not the case, $C(I, e)$ is produced.

Below we give the algorithms of Edmonds and Petersen. Both require as input a set I partitioned into subsets $I_1, \dots, I_k$ such that $I_j \in \mathcal{I}_j$, $j = 1, \dots, k$, and some $e \in E \setminus I$. Furthermore, independence testing subroutines for $M_1, \dots, M_k$ are assumed to be available as are routines for computing rk_i and sp_i, $i = 1, \dots, k$. No proofs of the validity of the algorithms are given, the reader is referred to [12] and [16].

Edmonds' algorithm [6]

Step 1: 1.0: $S_0 := I \cup \{e\}$; $j := 0$;

1.1: Find the smallest index i s.t.
$|I_i \cap S_j| < \mathrm{rk}_i(S_j)$;
if no such index exists STOP,
the set S_j satisfies $|S_j| > \sum_{i=1}^{k} \mathrm{rk}_i(S_j)$ and hence $I \cup \{e\}$ is not partitionable;

1.2: $S_{j+1} := S_j \cap \mathrm{sp}_i(I_i \cap S_j)$;
$m(j) := i$;

1.3: If $e \in S_{j+1}$ then $j := j + 1$; go to Step 1.1 else go to Step 2;

Step 2: 2.1: $I_{m(j)} := I_{m(j)} \cup \{e\}$;
If $I_{m(j)} \in \mathcal{I}_{m(j)}$ then go to Step 2.3;
2.2: Determine $C_{m(j)}(I_{m(j)}, e)$;
Choose $e' \in C_{m(j)}(I_{m(j)}, e) \setminus S_j$;
$I_{m(j)} := I_{m(j)} \setminus \{e'\}$;
$e := e'$; $j := j-1$;
go to Step 2.1;
2.3: STOP, $I_1, \dots, I_k$ is now a partition of $I \cup \{e\}$;

Petersen's algorithm [16]

Step 0; new $:= \{e\}$; add $:= \emptyset$; label$(e) := 0$;
for all $a \in I$ label$(a) := 0$;
Step 1: 1.0: if new $= \emptyset$ then
STOP, $I \cup \{e\}$ is not partitionable and
$C(I, e) = \{e\} \cup \{x \in I \mid \text{label}(x) \neq 0\}$;
1.1: for all $j \in \{1, \dots, k\}$ and all $a \in$ new do
if $I_j \cup \{a\} \in \mathcal{I}_j$ then go to Step 2
else for $b \in C_j(I, a)$ s.t. label$(b) = 0$ do
label$(b) := a$;
add $:=$ add $\cup \{b\}$;
new $:=$ add;
add $:= \emptyset$;
go to Step 1;
Step 2: 2.0: $x := a$; $lx :=$ label(a);
2.1: if $lx = 0$ then $I_j := I_j \cup \{e\}$ else
find $p \in \{1, \dots, k\}$ s.t. $x \in I_p$;
$I_j := I_j \cup \{x\}$; $I_p := I_p \setminus \{x\}$;
$x := lx$; $lx :=$ label(x); $j := p$;
2.2: STOP, $I_1, \dots, I_k$ is now a partition of $I \cup \{e\}$;

In Section 5 we specialize both algorithms to the case in which all of the k matroids are the circuit matroid of a given graph. However, a few comments on the complexity of the general algorithms are appropriate in advance.

It is easy to show that the worst-case complexity of both algorithms is $0(k \cdot |E|^2 \cdot f(|E|))$, where $f(|E|)$ is the maximum number of computational steps required to test independence in any of $M_1, \dots, M_k$. Edmonds' algorithm at first glance seems a bit more complicated due to the computations of rank functions and spans in $M_1, \dots, M_k$. These functions can, however, be implemented by means of the routines for computing fundamental circuits in each of the matroids. A point in favour of the algorithm is that in case efficient algorithms not involving computation of fundamental circuits are available only the circuits necessary to construct the extended partition are computed while the algorithm of Petersen generally computes a large number of others. The computational experiences reported in Section 6 confirm the practical importance of this point.

5. Solving the *k*-MSTP

Let $G=(V,E)$ be a connected graph with non-negative weights on the edges and consider the problem of finding k edge-disjoint spanning trees of G of minimum total weight (k-MSTP). This can be formulated as a minimum weight base problem of a sum matroid as follows: Defining $\mathcal{I}$ as the family of circuit-free subsets of E, the system $M=(E,I)$ is a matroid called the circuit matroid of G. Let $M_1=\cdots=M_k=M(G)$. A base of $M(G)$ is a spanning tree of G, and provided that k edge-disjoint spanning trees of G exist, a base of $M_1 \vee \cdots \vee M_k$ is the union of k such trees. Hence the k-MSTP is the problem of finding a minimum weight base in the matroid M. The algorithm described below solves the corresponding maximization problem, but as noted in Section 2 there is no difficulty in changing this to solve minimization problems (or change the problem to a maximization problem). In the algorithm I is the existing partial solution and U the edges of G not yet considered. For $e \in E$, $w(e)$ denotes the weight of e. Note that in case k edge-disjoint spanning trees do not exist this is indicated by the cardinality of the "solution" determined by the algorithm.

The greedy algorithm for the k-MSTP

Step 0: $I:=\emptyset$; $U:=E$;
Step 1: Select an edge $e \in U$ such that

$$w(e)=\max_{e' \in U}\{w(e')\};$$

Step 2: if $I \cup \{e\}$ is independent in $M=M_1 \vee \cdots \vee M_k$ then
$I:=I \cup \{e\}$;
if $|I|=k\cdot(n-1)$ then STOP, the existing partition $I=I_1 \cup \cdots \cup I_k$ solves the k-MSTP
else if $|U|=1$ then STOP, no solution exists;
$U:=U \setminus \{e\}$;
go to Step 1:

To implement this algorithm one has to decide on data structures to represent a graph, a forest etc. in a computer. However, to be reasonable this decision must take into account the characteristics of the graph in terms of e.g. number of edges. We have chosen to concentrate on large sparse graphs, and such a graph is represented economically (in terms of core storage) by the sequence of its edges. Furthermore, in the present case a weight is assigned to each edge and in each execution of Step 2, the edge of largest weight is to be selected. To facilitate this operation the graph is represented as a heap (cf. [1]).

In Step 2 of the algorithm the test for independence of $I \cup \{e\}$ is to be performed by either Edmonds' or Petersen's algorithm. Hence it is necessary to represent I in partitioned form, i.e. by $I_1, \ldots, I_k$. The representation of each I_j must furthermore enable construction of an efficient routine for computing $C(I, e)$ since such a routine is essential for the efficiency of especially Petersen's algorithm. Each I_j is a forest in G, and we have chosen to represent this as a collection of trees each of which is represented by means of the threaded index of Glover et al. [7]. This consists of two indices for each vertex of the (rooted) tree—one pointing to the "father" of the vertex and one pointing to the successor of the vertex in a preorder traversal of the tree. Since each vertex of G belongs to at most one of the trees constituting I_j this requires two indices per vertex for each I_j. The threaded index is well suited for determining whether $I \cup \{e\}$ is independent, and the removal and insertion of edges in the sets $I_1, \ldots, I_k$ can be implemented very effectively. A drawback, however, is that the height of the trees cannot be controlled, which means that the amount of work spent when searching in vain for a circuit $C(I_j, e)$ may be unnecessarily large.

While the computation of fundamental circuits is the only crucial point in Petersen's algorithm, an effective implementation of Edmonds' algorithm requires routines for computing rank functions and spans. The rank function, however, is easily dealt with in case of k identical matroids. Recall that in the jth execution of Step 1.1 of the algorithm, $\mathrm{rk}_i(S_j)$ is computed for each $i \in \{1, \ldots, k\}$. Now from Step 1.2

$$S_j = S_{j-1} \cap \mathrm{sp}_{i'}(I_{i'} \cap S_{j-1})$$

for some specific i'. But then for any $i \in \{1, \ldots, k\}$,

$$rk_i(S_j) = \mathrm{rk}_{i'}(S_j) = \mathrm{rk}_{i'}(I_{i'} \cap S_{j-1}) = |I_{i'} \cap S_{j-1}|,$$

and $|I_{i'} \cap S_{j-1}|$ is computed in the $(j-1)$th execution of Step 1.1. Hence only $\mathrm{rk}_i(S_0)$ has to be computed "explicitly", and since each matroid is graphic this is an easy task.

W.r.t. the computations of S_{j+1} in Step 1.2, the situation is somewhat more complicated. With going into detail we mention that $\mathrm{sp}_i(I_i \cap S_j)$ is computed by constructing the components of the forest $I_i \cap S_j$ from a list representation of S_j and the threaded-index representation of I_i. The initial list for S_0 is constructed by linking the components of each I_i together. Having determined the components of $I_i \cap S_j$, the elements of $S_j \setminus I_i$ are checked against these by traversing the S_j-list and excluding an edge (p, q) from this if p and q belong to different components of $I_i \cap S_j$.

6. Computational results

In this section we give some computational experience obtained by running the two implemented versions of the greedy algorithm on a UNIVAC 1100 computer. The algorithms have been programmed in the language PASCAL. The value of k has been set to 2 thus trying to get an idea of whether a solution of the PSP with $k=2$ may be based on a subroutine solving a sequence of k-MSTP-problems.

The first part of the analysis contains results obtained with some randomly generated graphs (in fact multigraphs) with from 10 to 150 vertices. The edges are generated in random order corresponding to ordering these by decreasing randomly generated edge weights. The number of edges in each graph is approximately one tenth of the number of edges in a complete graph with the same number of vertices. The graphs are of different structure in that some have many edges discarded during the solution and some only few. To reflect this, the results are given in two different ways. Fig. 2 shows the running times as a function of the number of vertices in the graph, and Fig. 3 shows the running times as a function of the number of edges actually chosen in Step 2 of the algorithm. It should be noted that due to limited resources of computer time the results stated are based on very few runs for each size of the graph. Hence the results give some "hints" about the behaviour of the algorithms rather than conclusive evidence on their effectiveness.

When the size of the graph increases, Edmonds' algorithm seems to perform much better than Petersen's. This is probably due to the structure of the two algorithms as they work quite differently. Petersen's algorithm tries to find an augmenting path even if no such path exists thus requiring a large amount of unnecessary computation to find fundamental circuits. Furthermore, all of these have to be inspected to discover unlabelled edges. Edmonds' algorithm on the other hand, determines in advance the existence/non-existence of an augmenting

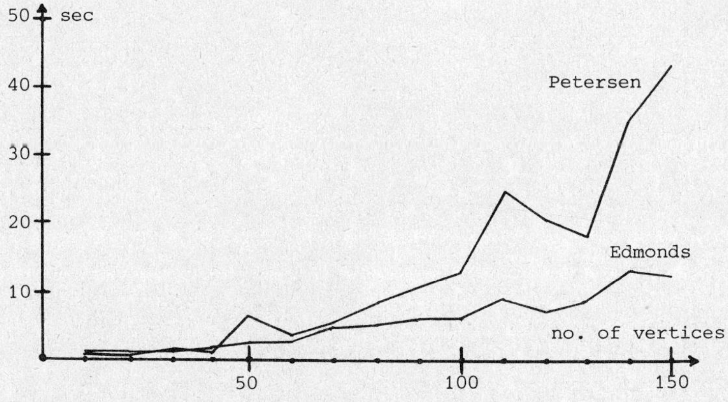

Fig. 2. Running time as a function of number of vertices in G.

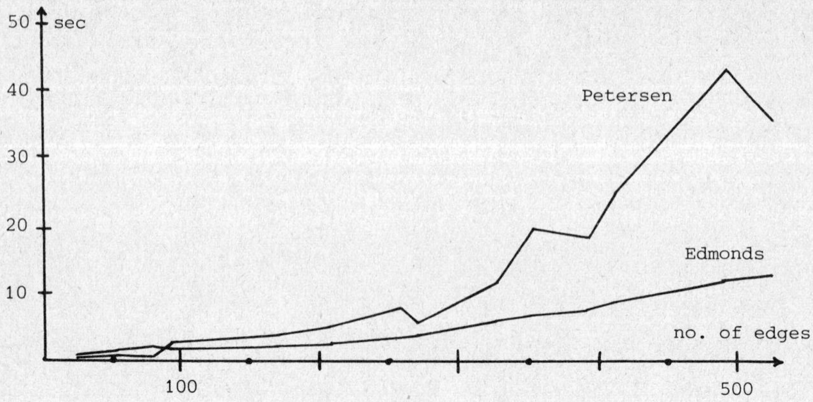

Fig. 3. Running time as a function of number of examined edges.

path, and only in case of an affirmative answer, the circuits determining the path are computed. To elucidate this we have calculated the length of the "augmenting" paths of Petersen's algorithm and of the sequence of S-set generated by Edmonds' algorithm in case it is not possible to fit the edge considered into the existing solution. The results are given in Table 1. It should be noted that our implementation of the algorithms discovers exactly the same augmenting path in case $I \cup \{e\}$ is independent. The computations were performed on graphs each having 100 vertices, but again of different structure.

Table 1

	Time	Number of discarded edges	Lengths of "augmenting" paths												
			1	2	3	4	5	6	7	8	9	10	11	12	13
Edmonds	5.0	44	—	24	20	—	—	—	—	—	—	—	—	—	—
Petersen	7.5		—	—	—	—	—	15	26	3	—	—	—	—	—
Edmonds	6.4	88	—	36	38	14	—	—	—	—	—	—	—	—	—
Petersen	10.8		—	—	—	—	—	6	23	36	20	3	—	—	—
Edmonds	7.1	137	—	116	13	8	—	—	—	—	—	—	—	—	—
Petersen	17.7		—	—	—	—	11	59	39	16	7	5	—	—	—
Edmonds	8.2	154	—	89	34	6	25	—	—	—	—	—	—	—	—
Petersen	13.5		—	—	—	—	15	37	41	24	18	16	2	—	1
Edmonds	7.8	199	97	43	38	21	—	—	—	—	—	—	—	—	—
Petersen	16.9		—	1	—	—	17	100	68	13	—	—	—	—	—

The results of Figs. 2 and 3 indicate that Edmonds' algorithm is superior to Petersen's for large graphs. Furthermore, Table 1 suggests that even a more efficient implementation of Petersen's algorithm should be used only in case $C(I, e)$ is essential for other computations. If a k-MSTP-routine is to be applied in solving the PSP, this routine should be based on Edmonds' partition algorithm. We have run this version for a few graphs with more than 150 vertices to indicate the size of the graphs tractable by the algorithm. The results are given in Table 2 and show that the algorithm is reasonably effective even for large graphs. In view of these results, it seems realistic to base a PSP-algorithm or heuristic on the Edmonds' version of the k-MSTP-algorithm.

Table 2

Number of vertices	Number of edges considered	sec
200	632	19
300	1305	60
400	1870	108

7. Problems in solving the PSP

In trying to develop either an optimal algorithm or a heuristic for the PSP even with $k=2$, several problems arise, a few of which we will briefly discuss.

First note that given a graph $G=(V,E)$ the system $M=(E,\mathcal{I})$ with $\mathcal{I}$ defined by

$$\mathcal{I}=\{I\subseteq E \mid I \text{ contains at most one circuit of } G\}$$

is a matroid (cf. e.g. [17]), and both 1-trees and Hamiltonian circuits are bases of this. Hence unions of edge-disjoint 1-trees resp. Hamiltonian circuits are bases of $M \vee M$.

For the TSP, the Held and Karp procedure exploits the fact that any base of M w.r.t. which each vertex is of degree 2 is a Hamiltonian circuit. However, the generalization of this statement is not true; a base of $M \vee M$, in which each vertex is of degree 4, is not necessarily the disjoint union of two Hamiltonian circuits. In Example 2 any Hamiltonian circuit of G must include both of the heavy edges, but G is the union of the 1-trees T_1 and T_2. However, note that these are generated from different "1-vertices". Is the desired property fulfilled if the 1-trees are generated from the same vertex? To facilitate a branching strategy similar to the one of Held and Karp, either this question must be answered affirmatively or an algorithm to test whether a base of $M \vee M$ is the union of two edge-disjoint Hamiltonian circuits must be developed. The latter may not be easy as e.g. 4-regular 4-connected non-Hamiltonian graphs exist [13].

When generalizing the "$\frac{3}{2}$-heuristic" two major problems turn up. For a given disjoint union of two spanning trees, the status of a vertex v in G may be one of the following:

(a) v is of even degree in both trees,
(b) v is of odd degree in both trees,
(c) v is of even degree in one tree and odd degree in the other.

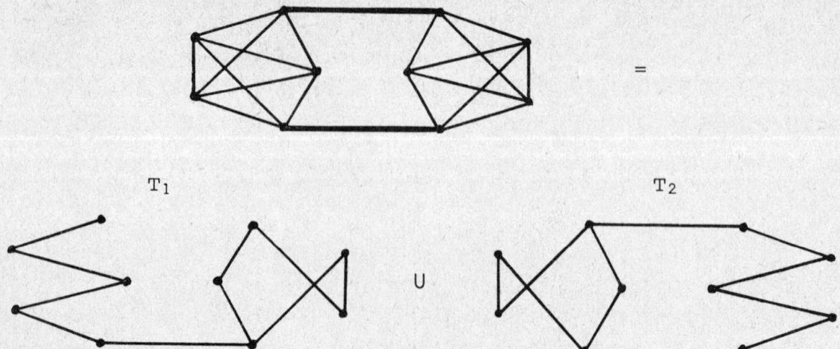

Example 2. A graph, which is the union of two edge-disjoint 1-trees, and in which two edge-disjoint Hamiltonian circuits do not exist.

If a minimum b-matching M_b (cf. Lawler [12]) for G is constructed such that the degree of a vertex of type a, b, and c in the matching is resp. 0, 2, and 1 we obtain a subgraph, in which each vertex is of even degree at least 4. Is a subgraph generated in this way always the disjoint union of two closed paths, each of which contains all vertices of G? If so, how can these be constructed?

By the procedure devised by Christofides, a Hamiltonian circuit is constructed from a path of this type. But how can we ensure that in case two circuits are constructed from edge-disjoint paths, these are disjoint?

Concluding we feel that in spite of the effective algorithm devised for the k-MSTP, the solution of the PSP both optimally and approximately is still far away, and further research in any of the open problems described above will be a step in the right direction.

References

[1] A.V. Aho, J.E. Hopcroft and J.D. Ullman, *The design and analysis of computer algorithms* (Addison-Wesley, Reading, MA, 1974).

[2] N. Christofides, "Worst-case analysis of a new heuristic for the travelling salesman problem", Management science report no. 388, Carnegie-Mellon University (1976).

[3] N. Christofides and C. Whitlock, "Graph connectivity and vulnerability, a survey", Manuscript presented at the summer school on combinatorial optimization, Urbino, Italy (1978).

[4] J. Clausen, "Matroids and combinatorial optimization", Report no. 78/4, Institute of Datology, University of Copenhagen, Denmark (1978).

[5] J. Clausen and T. Hoholdt, "On the sum of matroids", Research report, Institute of Mathematics, Technical University of Denmark (1975).

[6] J. Edmonds, "Minimum partition of a matroid into independent subsets", *Journal of the National Bureau of Standards* 69B (1965) 67–72.

[7] F. Glover, D. Klingman and J. Stuts, "Augmented threaded index method for network optimization", *Journal of Operational Research and Information Processing* 12 (1974) 293–298.

[8] K.H. Hansen and J. Krarup, "Improvements of the Held–Karp algorithm for the symmetric travelling salesman problem", *Mathematical Programming* 7 (1975) 87–96.

[9] M. Held and R.M. Karp, "The travelling salesman problem and minimum spanning trees: Part II", *Mathematical Programming* 1 (1971) 6–25.

[10] D. Knuth, "Matroid partitioning", Research report no. STAN-CS-73-342, Stanford University (1973).

[11] J. Krarup, "The peripatetic salesman and some related unsolved problems", in: B. Roy, ed., *Combinatorial programming: methods and applications* (D. Reidel Publishing Company, Dordrecht, 1975) pp. 173–178.

[12] E.L. Lawler, *Combinatorial optimization: networks and matroids* (Holt, Rinehart and Winston, New York, 1976).

[13] G.H.J. Meredith, "Regular n-valent n-connected non Hamiltonian non-edge-colorable graphs", *Journal of Combinatorial Theory* 14(B) (1973) 55–60.

[14] L. Mirsky, *Transversal theory* (Academic Press, London, 1971).

[15] B. Petersen, "Investigating solvability and complexity of linear active networks by means of matroids", Research report, Institute of Mathematics, Technical University of Denmark (1977).

[16] B. Petersen, "Circuits in the union of matroids: an algorithmic approach", Research report, Institute of Mathematics, Technical University of Denmark (1978).

[17] D.J.A. Welsh, *Matroid Theory* (Academic Press, London, 1976).

Mathematical Programming Study 13 (1980) 102–110.
North-Holland Publishing Company

THE DISTANCE BETWEEN NODES FOR A CLASS OF RECURSIVE TREES

J.S. CLOWES

University of Newcastle-upon-Tyne, Newcastle-upon-Tyne, Great Britain

Received 1 February 1980

An operation of addition is defined on the set of random trees and a class of random trees which may be constructed by repeated application of this process is studied. These trees are, rather distantly, related to those which occur in some algorithms for the classical assignment problem. It is shown that the mean path lengths for trees in the class is bounded above by $n^{\beta-1}$, where n is the order of the tree and $\beta = \log_2 3$. This bound is attained for certain trees whose order is a power of 2.

Key words: Assignment, Combinatorial, Maximum Moment, Path Lengths, Random Trees, Recursive, Transportation.

1. Introduction

Many algorithms for the solution of combinatorial problems involve the construction of rooted trees and in most cases the mean distance from the nodes to the root is a significant quantity. The varieties of binary rooted trees associated with algorithms from coding theory and for searching and sorting are well-known and have been extensively studied. Rooted trees also occur in some algorithms for the classical assignment and transportation problems (Adams [1]). In these algorithms rooted trees are constructed in two stages. First, a free tree is built by linking together smaller trees. Then, as a final step, one of the nodes is chosen as root. Thus, in this case, the quantity of interest is really the mean distance between nodes.

As a model of the above process we consider an operation which constructs trees by linking at random two smaller trees. The class of trees which may be generated by repetition of this operation generalizes the recursive random trees discussed by Moon [2]. Our main result states that for trees in this class the mean path length is bounded above by $n^{\beta-1}$, where n is the order of the tree and $\beta = \log_2 3$.

2. Notation and definitions

Let T be a tree of order n, that is a connected graph on n nodes without cycles. If i and j are nodes of T there is a unique path joining them, the length of this path is equal to the number of edges in it and will be denoted by l_{ij}.

For any node r on T we define $\lambda_r(T)$ by,

$$\lambda_r(T) = \frac{1}{n} \sum_{i \in T} l_{ir}$$

where the notation implies that the summation extends over all nodes of T. $\lambda_r(T)$ is the mean distance of the nodes of T from node r. The mean distance between all pairs of nodes, denoted by $l(T)$, may now be defined through,

$$l(T) = \frac{1}{n} \sum_{r \in T} \lambda_r(T).$$

In what follows we find it more convenient to consider, instead of $\lambda_r(T)$ and $l(T)$, the quantities $\mu_r(T)$ and $m(T)$ defined through,

$$\mu_r(T) = \sum_{r \in T} (l_{ir} + 1) \tag{2.1}$$

$$= n(\lambda_r(T) + 1),$$

$$m(T) = \frac{1}{n} \sum_{r \in T} \mu_r(T) \tag{2.2}$$

$$= n(l(T) + 1).$$

$m(T)$ will be called the "moment of T".

3. Random trees

A random tree $\mathcal{T}$ of order n is a function defined on a space of random events and taking values from the set of trees on a given set of n vertices. We denote by $\bar{m}(\mathcal{T})$ the mean moment of $\mathcal{T}$, that is, the expected value of the random variable $m(\mathcal{T})$.

A convenient method for defining a specific random tree is to describe an algorithm for constructing instances of it. Such an algorithm must provide, at least implicitly, a specification of the space of random events and of the mapping from this space onto the relevant set of trees.

Let $\mathcal{T}$ and $\mathcal{T}'$ be random trees on disjoint sets of order n and n' respectively. We define $\mathcal{T} \oplus \mathcal{T}'$ to be the random tree of order $n + n'$ generated by the following algorithm.

(1) Choose instances T and T' of $\mathcal{T}$ and $\mathcal{T}'$ respectively.

(2) Choose a node i on T and a node i' on T', all nodes having equal probability of being chosen.

(3) Join i and i'.

This operation of summation is commutative but not associative, since in the tree $(\mathcal{A} \oplus \mathcal{B}) \oplus \mathcal{C}$ the trees $\mathcal{A}$ and $\mathcal{B}$ are necessarily linked but this is not the case in $\mathcal{A}\oplus(\mathcal{B}\oplus\mathcal{C})$. We propose to study the values of the mean moments of random trees generated recursively by repeated summation. Our discussion will be based on the following theorem.

Theorem 1. *If $\mathcal{T}$ and $\mathcal{T}'$ are random trees of order n and n' respectively then,*

$$\bar{m}(\mathcal{T}\oplus\mathcal{T}') = \left(1+\frac{n'}{n+n'}\right)\bar{m}(\mathcal{T}) + \left(1+\frac{n}{n+n'}\right)\bar{m}(\mathcal{T}').$$

Proof. Let T and T' be instances of $\mathcal{T}$ and $\mathcal{T}'$ and let T_{ij} be the tree obtained by linking node i on T to node j on T'. The value of $\bar{m}(\mathcal{T}\oplus\mathcal{T}')$ will be computed by first calculating the mean of the moments of the $n\cdot n'$ trees T_{ij} and then averaging this over all pairs (T, T'), each weighted by its probability.

From (2.2) we have,

$$\frac{1}{nn'}\sum_{i\in T}\sum_{j\in T'} M(T_{ij}) = \frac{1}{nn'}\sum_{i\in T}\sum_{j\in T'}\left\{\frac{1}{n+n'}\left(\sum_{r\in T}\mu_r(T_{ij}) + \sum_{r\in T'}\mu_r(T_{ij})\right)\right\}.$$

Consider the contribution to this sum of the terms with r in T. In this case the distance of a node of T' from r is equal to its distance from j plus $l_{ir}+1$, the 1 arising because of the edge linking i and j. Thus, using (2.1), we find,

$$\mu_r(T_{ij}) = \mu_r(T) + n'\cdot(l_{ir}+1) + \mu_j(T')$$

whence

$$\begin{aligned}\sum_{i\in T}\sum_{j\in T'}\sum_{r\in T}\mu_r(T_{ij}) &= nn'\sum_{r\in T}\mu_r(T) + (n')^2\sum_{r\in T}\sum_{i\in T}(l_{ir}+1) + n^2\sum_{j\in T'}\mu_j(T') \\ &= n^2n'\, m(T) + (n')^2\sum_{r\in T}\mu_r(T) + n^2n'\, m(T) \\ &= nn'(n+n')\cdot m(T) + n^2n'\cdot m(T').\end{aligned}$$

Similarly, the terms with r in T' contribute an amount equal to

$$n(n')^2\cdot m(T) + nn'(n+n')\cdot m(T').$$

Upon adding these two expressions and dividing by $nn'(n+n')$ we find the mean of the moments of the nn' trees T_{ij} to be equal to,

$$\left(1+\frac{n'}{n+n'}\right)\cdot m(T) + \left(1+\frac{n}{n+n'}\right)\cdot m(T').$$

Averaging this expression over all pairs (T, T') yields the required result.

4. Examples of recursive random trees

For $n = 1, 2, \ldots$, let $\mathcal{A}_n$ be the random tree of order n defined by the recurrence relations,

$\mathcal{A}_1$ is the tree of order 1.

$\mathcal{A}_n = \mathcal{A}_{n-1}\oplus\mathcal{A}_1, \quad n = 2, 3, \ldots .$

Roughly speaking, $\mathscr{A}_n$ is the sum of a sequence of n individual nodes and is the recursive tree considered by Moon [2].

If we denote the mean moment of $\mathscr{A}_n$ by a_n, then,

$$a_1 = 1,$$

$$a_n = \left(1+\frac{1}{n}\right)a_{n-1} + \left(1+\frac{n-1}{n}\right)$$

$$= \frac{2n-1}{n} + \frac{n+1}{n}\left(\frac{2n-3}{n-1} + \frac{n}{n-1}a_{n-2}\right)$$

$$= (n+1)\left\{\frac{2n-1}{(n+1)\cdot n} + \frac{2n-3}{n\cdot(n-1)} + \cdots + \frac{1}{2\cdot 1}\right\}$$

$$= (n+1)\sum_{k=1}\left(\frac{3}{k+1} - \frac{1}{k}\right)$$

$$= 2(n+1)H_n - 3n$$

where $H_n = 1 + 2^{-1} + 3^{-1} + \cdots + n^{-1}$. This agrees with one of Moon's results.

We may usefully generalize the above construction as follows. Let $\mathscr{T}$ be a random tree of order t. For $n = 1, 2, \ldots$, let $\mathscr{A}_n(\mathscr{T})$ be the random tree of order nt defined by

$$\mathscr{A}_1(\mathscr{T}) = \mathscr{T},$$

$$\mathscr{A}_n(\mathscr{T}) = \mathscr{A}_{n-1}(\mathscr{T}) \oplus \mathscr{T}, \quad n = 2, 3, \ldots .$$

Thus $\mathscr{A}_n(\mathscr{T})$ is the sum of a sequence of n replicates of $\mathscr{T}$. Denoting the mean moment of $\mathscr{A}_n(\mathscr{T})$ by $a_n(\mathscr{T})$ we have,

$$a_1(\mathscr{T}) = \bar{m}(\mathscr{T}),$$

$$a_n(\mathscr{T}) = \left(1+\frac{1}{n}\right)\cdot a_{n-1}(\mathscr{T}) + \left(1+\frac{n-1}{n}\right)\cdot \bar{m}(\mathscr{T})$$

whence

$$a_n(\mathscr{T}) = a_n \cdot \bar{m}(\mathscr{T}).$$

This result suggests that for random trees constructed by linking at random a sequence of components of fixed size, the mean path length increases only as the logarithm of the order of the tree.

Now, let $r \geq 2$ be a fixed integer and for $k = 0, 1, 2, \ldots$, let $\mathscr{B}_k(r)$ be the random tree of order $n = r^k$, defined by

$\mathscr{B}_0(r)$ is the tree of order 1.

$$\mathscr{B}_k(r) = \mathscr{A}_r(\mathscr{B}_{k-1}(r)).$$

That is $\mathscr{B}_k(r)$ is the sum of r replicates of $\mathscr{B}_{k-1}(r)$. For $b_k(r)$, the mean moment of $\mathscr{B}_k(r)$, we have by our previous result,

$$b_k(r) = a_r \cdot b_{k-1}(r) = a_r^k = a_r^{\log_r n} = n^{\log_r a_r}.$$

Since $a_r > r$, for trees constructed in this way the mean moment increases as some power of n greater that the first. In particular, for $r = 2$, the tree $\mathcal{B}_k(2)$ of order $n = 2^k$ has mean moment $n^{\log_2 3} \doteqdot n^{1.58}$ and the mean path length increases as $n^{0.58}$.

A final example serves to illustrate a more general method for defining recursive random trees. We denote by $\mathcal{C}_n$ the random tree of order n defined by

$\mathcal{C}_1$ is the tree of order 1.

For $n \geq 2$ instances of $\mathcal{C}_n$ are generated by the following algorithm.

Step 1: Choose an integer k in the range $1 \leq k \leq n-1$, all integers in the range being equally probable.

Step 2: Choose an instance of $\mathcal{C}_k \oplus \mathcal{C}_{n-k}$.

The mean moment c_n of $\mathcal{C}_n$ satisfies,

$$c_1 = 1,$$

$$c_n = \frac{1}{n-1} \sum_{k=1}^{n-1} \bar{m}(\mathcal{C}_k \oplus \mathcal{C}_{n-k}) \quad n = 2, 3, \ldots$$

$$= \frac{2}{n-1} \sum_{k=1}^{n-1} \left(1 + \frac{n-k}{n}\right) c_k.$$

This recurrence relation seems to have no simple analytic solution. However, it is possible to show that, for large n, c_n is proportional to $n^{\gamma-1}$, where $\gamma \doteqdot 2.56$ is the positive root of the equation

$$\gamma^2 - \gamma - 4 = 0.$$

5. Random trees with maximum moment

The results of the last section show that the range of possible values for the mean moments of recursive random trees generated by the summation operation is very large. We seek now to determine the limits of this range. In this section we obtain an upper bound for the mean moment and discuss the problem of how random trees with maximum moment may be generated.

First it is necessary to define more precisely the class of random trees to be considered. We call this the class of admissible trees and define it, recursively, as follows.

Definition. (1) The tree of order 1 is admissible.

(2) A random tree of order n is admissible if, and only if, it is definable by an algorithm whose effect is to construct with some specific probability $\pi(\mathcal{T}, \mathcal{T}')$ an instance of the random tree $\mathcal{T} \oplus \mathcal{T}'$, where $(\mathcal{T}, \mathcal{T}')$ is a pair of admissible trees whose orders sum to n.

It is easy to show that the trees $\mathscr{A}_n$, $\mathscr{B}_k(r)$ and $\mathscr{C}_n$ defined in the last section are all admissible. Also, it is evident that the sum of any two admissible trees is itself admissible and that the mean moment of the most general kind of admissible tree is a weighted average of the moments of trees of this restricted type. Thus, the maximum value for the mean moment of an admissible tree of order n must be attained for at least one tree which is simply the sum of two smaller admissible trees.

Let w_n be the maximum moment for an admissible tree or order n and let $\mathscr{T}$ and $\mathscr{T}'$ be trees of order k and $n-k$ respectively, $1 \le k \le n-1$. By Theorem 1,

$$\bar{m}(\mathscr{T} \oplus \mathscr{T}') = \left(1 + \frac{n-k}{n}\right)\bar{m}(\mathscr{T}) + \left(1 + \frac{k}{n}\right)\bar{m}(\mathscr{T}').$$

For fixed k and n the maximum value of this expression as $\mathscr{T}$ and $\mathscr{T}'$ range independently over the appropriate sets of admissible trees is,

$$\left(1 + \frac{n-k}{n}\right) w_k + \left(1 + \frac{k}{n}\right) w_{n-k}.$$

To obtain w_n we must choose k to maximise this expression. We have proved:

Theorem 2. *For $n = 1, 2, \dots$, the maximum moment for an admissible tree of order n is the solution of the recurrence relations,*

$$w_1 = 1,$$

$$w_n = \max_{k=1,2,\dots,n-1} \left\{ \left(1 + \frac{n-k}{n}\right) w_k + \left(1 + \frac{k}{n}\right) w_{n-k} \right\}, \quad n = 2, 3, \dots .$$

Given w_i for $i = 1, 2, \dots, n-1$ we can compute w_n from the above relation in $O(n)$ operations. At the same time we determine an optimal value $k_{\max}$ for k and so can describe how to construct an admissible tree with moment w_n.

The determination of an analytic solution to the recurrence relation is more difficult. In this direction we have the following theorem, which gives only an upper bound. Since the bound is attained whenever $n = 2^k$ by the tree $\mathscr{B}_k(2)$ it is, in a sense, best possible.

Theorem 3.

$$w_n \le n^\beta, \quad n = 1, 2, 3, \dots$$

where

$$\beta = \log_2 3$$

Proof. Consider the function $y(x)$ defined for all real $x \ge 1$ by

$$y(x) = x^\beta, \quad 1 \le x \le 2,$$

$$y(x) = \max_{1 \le t \le x-1} \left\{ \left(1 + \frac{x-t}{x}\right) y(t) + \left(1 + \frac{t}{x}\right) y(x-t) \right\}, \quad x \ge 2. \tag{5.1}$$

If t was required to be integral in (5.1) $y(x)$ would be equal to w_n for integer values of x. Since the imposition of additional constraints cannot increase a constrained maximum we infer,

$$w_n \leq y(n), \quad n = 1, 2, \ldots . \tag{5.2}$$

We now assert that the solution of Eq. (5.1) is,

$$y(x) = x^\beta, \quad x \geq 1. \tag{5.3}$$

The proof is by induction on integral values of x. The assertion is certainly true for x in $1 \leq x \leq 2$, assume it true for x in some range $1 \leq x \leq n$, $n \geq 2$. Then for x in $n \leq x \leq n+1$,

$$y(x) = \max_{1 \leq t \leq x-1} \left\{ \left(1 + \frac{x-t}{x}\right) t^\beta + \left(1 + \frac{t}{x}\right)(x-t)^\beta \right\}. \tag{5.4}$$

For fixed x the expression on the right in (5.4) is symmetric about $t = \frac{1}{2}x$ so we may restrict t to the range $\frac{1}{2}x \leq t \leq x-1$. The substitution $t = \frac{1}{2}x(1+p)$ maps this onto $0 \leq p \leq (x-2)/x$ and (5.4) becomes,

$$y(x) = \frac{1}{2}\left(\frac{x}{2}\right)^\beta \max_{0 \leq p \leq (x-2)/x} \{(3-p)(1+p)^\beta + (3+p)(1-p)^\beta\}. \tag{5.5}$$

Thus we are concerned with the maximum value of the differentiable function

$$F(p) = (3-p)(1+p)^\beta + (3+p)(1-p)^\beta$$

on the interval $0 \leq p \leq (x-2)/x < 1$. Such a maximum must occur either at an endpoint of the range or at an interior stationary point. Since $F(0) = F(1) = 6$, there must be at least one stationary point in the (extended) interval $0 \leq p \leq 1$.

Differentiating with respect to p,

$$\begin{aligned} F'(p) &= \beta\{(3-p)(1+p)^{\beta-1} - (3+p)(1-p)^{\beta-1}\} - \{(1+p)^\beta - (1-p)^\beta\} \\ &= 4\beta\{(1+p)^{\beta-1} - (1-p)^{\beta-1}\} - (\beta+1)\{(1+p)^\beta - (1-p)^\beta\}. \end{aligned}$$

This vanishes when $p = 0$ and examination of $F''(0)$ shows F to have a maximum at this point.

Differentiating $(1+p)^{\beta-1} - (1-p)^{\beta-1}$ twice we find,

$$\frac{\mathrm{d}^2}{\mathrm{d}p^2}\{(1+p)^{\beta-1} - (1-p)^{\beta-1}\} = (\beta-1)(\beta-2)\{(1+p)^{\beta-3} - (1-p)^{\beta-3}\}$$

which is positive for $0 < p \leq 1$ since $1 < \beta < 2$. Thus, $(1+p)^{\beta-1} - (1-p)^{\beta-1}$ is "concave upwards" on $0 \leq p \leq 1$, that is, the chord joining any two points lies above the curve. A similar calculation shows $(1+p)^\beta - (1-p)^\beta$ to be concave downwards. It follows that $F'(p)$ can have only two zeros in $0 \leq p \leq 1$, and since $p = 0$ corresponds to a maximum value of F the other, interior, zero must correspond to a minimum.

We have proved that the maximum value of $F(p)$ on $0 \leq p \leq (x-2)/x$ is

$F(0)=6$. Substituting this value into (5.5) yields,

$$y(x)=\tfrac{1}{2}\cdot(\tfrac{1}{2}x)^{\beta}\cdot 6=x^{\beta}.$$

This proves the assertion (5.3) for $n\le x\le n+1$ and therefore, by induction, for all x. The statement of the theorem now follows from (5.2).

The proof of Theorem 3 suggests that the appropriate value for k in the recurrence relation for w_n is $k=\lceil\frac{1}{2}n\rceil$. The sequence u_n generated by this rule satisfies,

$$u_1=1,$$

$$u_n=\left(1+\frac{\lceil\frac{1}{2}n\rceil}{n}\right)u_{\lfloor n/2\rfloor}+\left(1+\frac{\lfloor\frac{1}{2}n\rfloor}{n}\right)u_{\lceil n/2\rceil},\quad n=2,3,\ldots$$

and the corresponding random trees are defined by,

$\mathcal{U}_1$ is the tree of order 1,

$$\mathcal{U}_n=\mathcal{U}_{\lceil n/2\rceil}\oplus\mathcal{U}_{\lfloor n/2\rfloor}\quad n=2,3,\ldots.$$

Computer calculations confirm that the trees $\mathcal{U}_n$ do have maximum moment for $n\le 80$, at least. The calculations also show that these are not the only admissible trees with maximum moment. For example, both $\mathcal{U}_3+\mathcal{U}_3$ and $\mathcal{U}_2+\mathcal{U}_4$ have moment 17. This lack of uniqueness is especially surprising in view of the results of Hammersley and Grimmett [3] who have studied a recurrence relation similar to that for w_n and have shown that, under very general conditions, $k=\lceil\frac{1}{2}n\rceil$ is the unique optimal value of k for their equation.

6. Conclusion

We have shown that for admissible trees of order n the mean path length may vary from approximately $3\log(n)$ to $n^{0.58}$. Thus, if the resource requirements of an algorithm depend upon the mean path length in such a tree we require very precise information about the method of construction of the tree in order to perform an exact analysis of the algorithm. Of course, the above-mentioned limits enable us to compute bounds for the requirements but they leave open a very wide range of variability.

Acknowledgment

I wish to thank the referee for many helpful comments and, in particular, for drawing my attention to reference [3].

References

[1] E. Adams, "An investigation into the use of records and references in the solution of the transportation problem", M.Sc. Disseration, University of Newcastle-upon-Tyne (1974).

[2] J.W. Moon, "The distance between nodes in recursive trees", in: T.P. McDonough and V.C. Marron, eds., *Combinatorics* (London Mathematical Society Lecture Notes, Series 13) London Mathematical Society, London, 1974) pp. 125–132.

[3] J.M. Hammersley and G.R. Grimmett, "Maximal solutions of the generalized subadditive inequality", in: E.F. Harding and D.G. Kendall, eds., *Stochastic geometry* (Wiley, London) pp. 270–284.

Mathematical Programming Study 13 (1980) 111–120.
North-Holland Publishing Company

OPTIMIZATION PROBLEMS ARISING FROM THE INCORPORATION OF SPLIT VALUES IN SEARCH TREES

V.J. RAYWARD-SMITH

University of East Anglia, Norwich, Great Britain

Received 1 February 1980

A review of techniques for the construction of optimal binary search trees leads to a new algorithm for the construction of nearly optimal split trees. Results are given to compare the efficiency of such trees with that of median split trees and some open problems arising are discussed. The concept of a generalised tree is introduced together with a description of how the solution of the difficult problems arising would have a considerable practical impact in the design of data bases.

Key words: Algorithm Analysis, B-trees, Data Base, Dynamic, Generalised Trees, Median Split Trees, Optimal Binary Search Trees, Split Trees.

1. Optimal binary search trees

Consider a set $K = \{k_1, k_2, \dots, k_n\}$ of keys under a well-ordering, $<$. A *binary search tree* for K is empty if $K = \emptyset$ or is a triple (root, left subtree, right subtree) where root $\in K$, left subtree is a binary search tree for $\{k \mid k \in K \text{ and } k < \text{root}\}$ and right subtree is a binary search tree for $\{k \mid k \in K \text{ and } k > \text{root}\}$. We will adopt the usual conventions for depicting binary search trees. As an example, Fig. 1 illustrates a binary search tree for 31 most common words in English (using alphabetic ordering). The amount of effort required to find whether a word (key) is in the tree clearly depends upon the level at which it occurs.

In a general situation, not every key will be equally likely to occur as a search argument. Using the notation of Knuth [10], let p_j denote the probability that k_j is a search argument and q_j denote the probability that the search argument lies

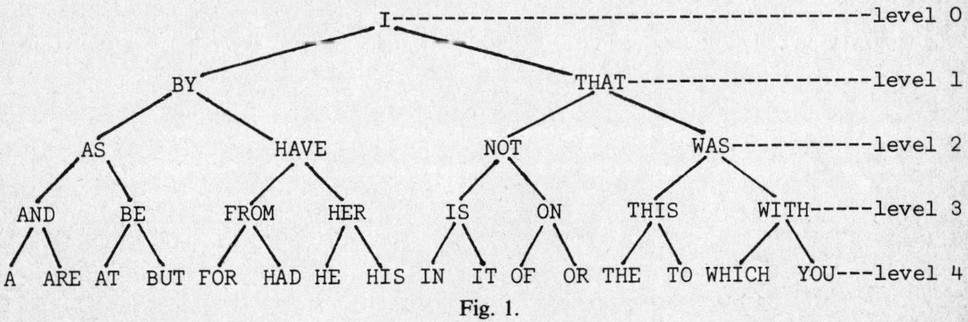

Fig. 1.

between k_j and k_{j+1} (by convention q_0 is the probability that the search argument is less than k_1, q_n is the probability that it is greater than k_n). Thus, $p_1 + p_2 + \cdots + p_n + q_0 + q_1 + \cdots + q_n = 1$. The expected number of comparisons in a search will be

$$\sum_{1 \le j \le n} p_j \text{ (level of } k_j + 1) + \sum_{0 \le k \le n} q_k \text{ (level of } (k+1)\text{th external node)}.$$

This will be called the *cost* of a search tree. A binary search tree of minimum cost will be called an *optimum search tree.* For example, given frequencies of the 31 common words as given in Table 1 and assuming each $q_k = 0$, the optimum search tree given in Fig. 2 has a cost of 3.437 (see [10, p. 433]).

Table 1
The frequencies of the 31 most common English words taken from Gaines [5]

A	5074	I	2292
AND	7638	IN	4312
ARE	1222	IS	2509
AS	1853	IT	2255
AT	1053	NOT	1496
BE	1535	OF	9767
BUT	1379	ON	1155
BY	1392	OR	1101
FOR	1869	THAT	3017
FROM	1039	THE	15568
HAD	1062	THIS	1021
HAVE	1344	TO	5739
HE	1727	WAS	1761
HER	1093	WHICH	1291
HIS	1732	WITH	1849
		YOU	1336

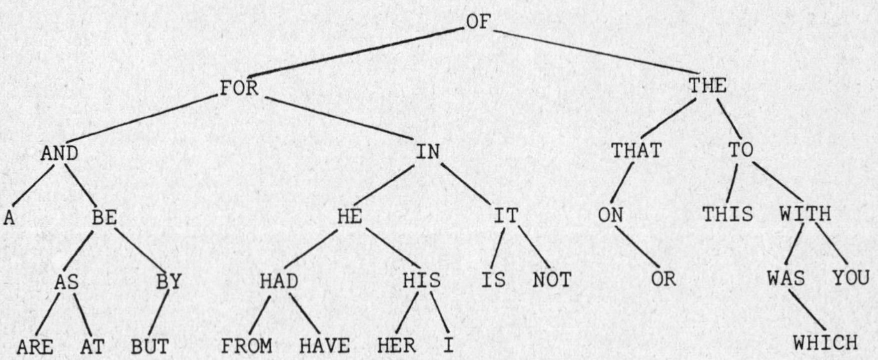

Fig. 2.

An algorithm for the construction of such a tree is described by Knuth in [8]. This algorithm is a clever modification of dynamic programming and requires $O(n^2)$ storage and $O(n^2)$ time. This is prohibitive for most applications and thus algorithms are used which take less space and time but construct only nearly optimal search trees. In Melhorn [11] there is an analysis of two of these which can be described as "place the most frequently occurring key at root of tree, then proceed similarly on the subtrees" and "choose the root so as to equalize as near as possible, the total weights of the left and right subtrees, then proceed similarly on the subtrees". Melhorn shows that the latter rule will always produce a good estimate. He outlines an implementation which requires $O(n \log n)$ time and $O(n)$ space, although Fredman [4] describes an implementation which only requires $O(n)$ time and $O(n)$ space. If the task is to minimise only the cost of unsuccessful searches, then a suitable tree can be constructed using the TC Algorithm described in Hu and Tucker [7] which also requires $O(n \log n)$ time and $O(n)$ space. A brief survey of these and other results is given in Hu [6].

2. Split trees

The key value in a binary search tree plays two important rôles. Firstly, it identifies the record which resides at that node and secondly, it partitions the remaining nodes between left and right subtrees. A *split tree* recognises these two rôles by having two distinct values stored in each node—a node-value which is the identifying key and a split-value which partitions the remaining nodes between left and right subtrees.

Procedure *insert* and *delete* given below can be used to insert and delete keys from split trees; *insert* will always insert a new node as a leaf node.

```
insert (key: k, tree: t) tree
   = if t is empty then tree val (node-value: k,
                                   split-value: k,
                                   left-subtree: empty,
                                   right-subtree: empty)
     else if k ≤ split-value of t then insert (k, left-subtree of t)
                                  else insert (k, right-subtree of t)
        fi
     fi
delete (key: k, tree: t) tree
 = if t isnt empty then
     s ← split-value of t; n ← node-value of t;
     lt ← left-subtree of t; rt ← right-subtree of t;
     ln ← node-value of lt;
```

```
if k = n then
tree val (node-value: ln,
          split-value: s,
          left-subtree: if ln ≠ s then delete (ln, lt)
                        else left-subtree of lt fi,
          right-subtree: rt)
  {optionally elsf k = s then
     tree val (node-value: n
               split-value: largest node-value in lt < s,
               left-subtree: lt,
               right-subtree: rt)}
  else tree val (node-value: n,
               split-value: s,
               left-subtree:  if k ≤ s then delete (k, lt)
                                       else lt fi,
               right-subtree: if k > s then delete (k, rt)
                                       else rt fi)
  fi
fi
```

The use of a second key in each node allows the most frequently occurring key to occur at the root without unduly affecting the structure of the remaining tree. In Shiel [12], median split trees are described; these are split trees where the root of each tree is so chosen and the split value is the median of the remaining keys with respect to the given ordering. Fig. 3 shows the median split tree for the 31 most common English words, again constructed using the frequencies given in [5]. The average cost per successful search is 3.127. (The split value of each node is shown in parentheses. By convention the largest key

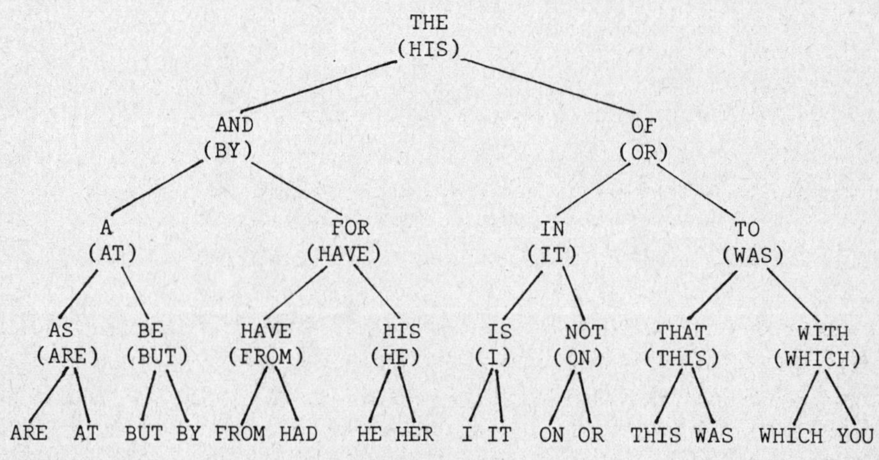

Fig. 3.

in the left-subtree will be used as a split value and thus split values can be omitted from such diagrams.)

The algorithm *insert* can also be adapted to use medians. The present version, having inserted a key k as the key-value at a leaf node, also uses k as the split-value. A better choice of split-value would be the median of the node-values which could be rooted at that subtree. This increases the complexity of the *insert* algorithm but might prevent unnecessarily long paths being developed.

The median split tree is attractive because it yields a balanced tree which can be efficiently stored [9, p. 401]. However, it is not the split tree of minimal cost. This is specially true if any q_k is significant but even if $q_k = 0$ for $k = 0, 1, \dots, n$, considerable improvements can be made. Shiel [12] claims that the determination of a split value for the optimum split tree is computationally intractable. This appears to be correct—certainly a generalisation of Knuth's $O(n^2)$ algorithm seems very difficult. However, it is easy to generalise the second algorithm discussed by Melhorn to obtain a near optimum split tree. The resulting algorithm is:

build $(k_1, \dots, k_n)$
  $=$ **if** $n = 0$ **then** empty
  else $v \leftarrow$ any k_i s.t. frequency$(k_i) \geq$ frequency(k_j) for $1 \leq j \leq n$;
  $s \leftarrow k_j$ such that $|\sum_{k \in Lk_j}$ frequency$(k) - \sum_{k \in Gk_j}$ frequency$(k)|$ is a minimum
    where $Lk_j = \{k \mid k \leq k_j$ and $k \neq k_i\}$
    and $Gk_j = \{k \mid k > k_j$ and $k \neq k_i\}$;
  tree **val** (node-value: v,
        split-value: s,
        left-subtree: build (Ls),
        right-subtree: build (Gs))
  fi

A similar argument to that given in Melhorn [11] can be used to show that this algorithm can be implemented in $O(n \log n)$ time which is of the same order as the algorithm used to construct median split trees. If *build* is applied to the 31 most common words example, the split tree given in Fig. 4 is constructed. The average cost per successful search is 3.137 which is $\frac{1}{3}\%$ worse than the cost for the median split tree. This is not too discouraging, since a 31 node example can be stored as a full binary tree of level four and thus might be expected to most favour the median split tree approach. In Table 2, results are given which compare the cost of near optimal split trees (NOSTs) with the costs of median split trees (MSTs). It can be seen that the difference in all cases is relatively insignificant but that NOSTs appear to perform a little better. For the randomly generated data we considered, the maximum depth of a NOST never exceeded the depth of a corresponding MST by more than one.

In the results given in Table 2, the distribution of the frequencies of nodes is

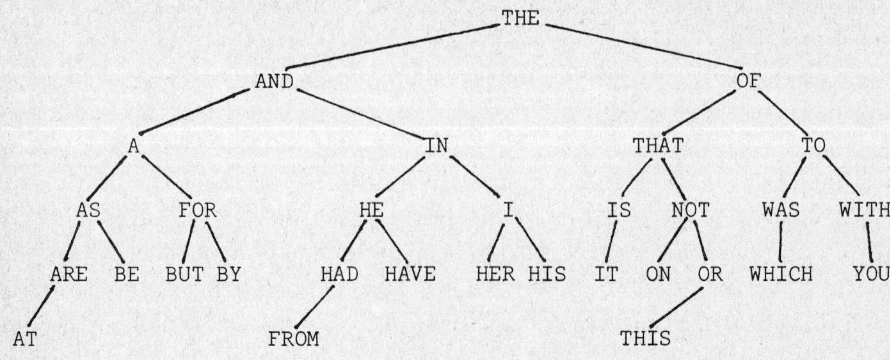

Fig. 4.

independent of the ordering of the nodes. However, in many practical applications this is not true. For example, a telephone directory will usually have more entries starting with S than any other letter. It is in such cases that NOSTs prove their real worth. As an experiment, we took the frequencies we had generated to construct the third column of Table 2. Rather than assign these randomly to the keys, we assigned the largest frequency generated to the least key (in lexicographic ordering), the second largest frequency generated to the second least key etc. This resulted in a situation where lower the key was in the lexicographic ordering the more likely it was to occur as a search argument. The comparison between the NOSTs and MSTs, given in Table 3, shows clearly the danger of using median split trees.

In many practical situations, the frequencies of the node-values in a split tree are difficult to predetermine. An initial guess at the frequencies may be possible but these should be updated each time the node-value occurs as a search argument. In such an environment, the split tree will be called *dynamic*.

There are two problems arising when dealing with dynamic split trees. The first is that the frequency of a node-value may be increased to be larger than that

Table 2
In all cases, integer frequencies were generated randomly in the range 1–1000
Average cost per successful search

Distribution of frequencies / Number of nodes	Uniform		Normal		Negative exponential	
	NOST	MST	NOST	MST	NOST	MST
10	2.270	2.353	2.297	2.359	2.151	2.156
20	3.061	3.129	3.201	3.249	2.890	2.923
30	3.553	3.613	3.749	3.755	3.172	3.195
40	3.967	4.020	3.989	3.916	3.377	3.410
50	4.230	4.275	4.417	4.440	3.767	3.819
100	5.160	5.193	5.338	5.355	4.620	4.666

Table 3

Number of nodes	Average cost per successful search	
	NOST	MST
10	2.196	2.408
20	2.940	3.231
30	3.444	3.663
40	3.590	3.949
50	4.026	4.337
100	4.880	5.314

of its parent and the second is that the tree may become overly left or right heavy. Ideally, we would like an efficient algorithm which kept a NOST "tuned", i.e., as the frequencies changed, the tree is slightly altered so that it remains near-optimal. Since the construction algorithm for a NOST is only $O(n \log n)$, efficient algorithms for "tuning" are difficult to find. Such algorithms have been discovered for ordinary search trees by Bruno and Coffman [2] and if they could be developed for NOSTs, they could prove very important. A general approach to dynamic trees where every time a node-value is accessed, it is interchanged with its parent (if possible) would result in commonly accessed nodes automatically filtering their way towards the root of the tree. The difficulty is that a simple interchange is only possible half of the time, i.e. either when the node appears in the left subtree and the parent node has a value less than or equal to its split value or when the node appears in the right subtree and the parent node has a value greater than its split value. If this technique is adopted, it is not surprising that we find any significant change in the frequencies can seriously degrade the tree. A better method might be to associate two node-values with each node, one node-value being greater than or equal to the split-value and one being less. Then interchanging a node-value with a node-value of the parent is always possible and the tree can be efficiently tuned. The drawback is that such trees are difficult to search. The general case is considered in Section 4.

3. B-trees

B-trees can be regarded as a generalisation of binary search trees which have been receiving considerable attention as a storage structure for certain files on paged secondary storage devices. A good survey is given in Comer [3]. In Bayer and McCreigh [1], a B-tree is defined as follows: Let $n \geq 0$ be an integer, m a natural number. A directed tree T is in the *class* $\tau(m, n)$ *of B-trees* if T is either empty ($n = 0$) or has the following properties:

(i) Each path from the root to any leaf has the same length n, also called the *height* of T, i.e. n = number of nodes in path.

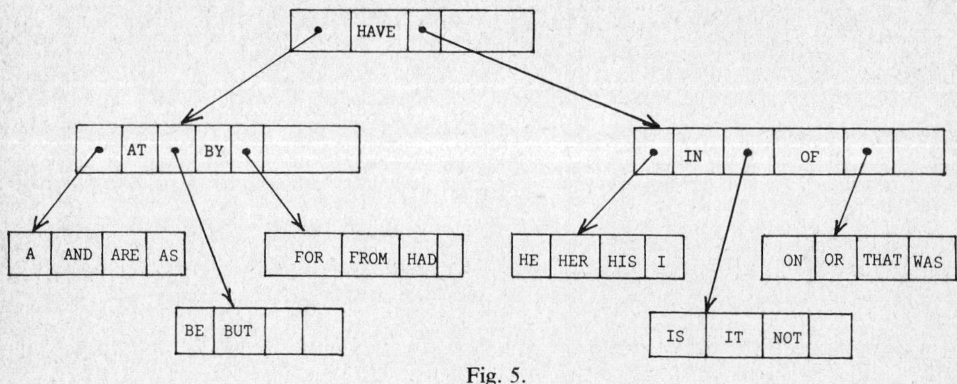

Fig. 5.

(ii) Each node except the root and the leaves has at least $m+1$ sons. The root is a leaf or has at least two sons.

(iii) Each node has at most $2m+1$ sons.

An example of a B-tree in the class $\tau(2,3)$ is given in Fig. 5. The keys are alphabetic and arranged so that within each node (assumed stored in one page) they are in increasing lexicographic order, $k_1, k_2, \ldots, k_l$; $m \le l \le 2m$, except for the root node for which $1 \le l \le 2m$. Furthermore, every non-terminal node contains $l+1$ pointers (references) $r_0, r_1, \ldots, r_l$ to the sons of the node. B-trees generally also satisfy an ordered property i.e. the keys, $K(r_i)$, in the subtree pointed to by pointer r_i are such that

$$\begin{aligned}&(\forall k \in K(r_0))(k < k_1),\\ &(\forall k \in K(r_i))(k_i < k < k_{i+1}), \quad i = 1, 2, \ldots, l-1,\\ &(\forall k \in K(r_l))(k_l < k).\end{aligned}$$

Ordered B-trees are important because retrieval, insertion and deletion of keys can all be achieved in time proportional to $\log_m |K|$ where $|K|$ is the number of keys. A full description and analysis of these algorithms is given in [1].

If the frequency of access of the various nodes alter, then there clearly exists an optimal B-tree of given class for given data of given frequencies. Since B-trees are designed particularly for simple insertion and deletion it is important that algorithms be found to not only generate a (near) optimal B-tree but also to update it. This is still an open problem. Ideally, one would like a dynamic situation where commonly accessed keys filter their way to the top of the tree. The 80–20 rule (80% of the keys are accessed 20% of the time) is a well-known observation of data base designers. In many data bases, there is an even more pronounced preference for an even smaller percentage of the keys.

4. Generalised trees

One can generalise the two concepts met in Sections 2 and 3. In Section 2, the distinction between a *split-value* and a *node-value* was made and in Section 3, the fundamental idea of having several keys associated with a node was introduced.

A *generalised tree* (G-tree) is designed to incorporate both ideas. A node in a G-tree can hold

(1) node-values (i.e. keys + associated records);
(2) split-values (simply keys);
(3) pointers (references) to subtrees.

The amount of store required for a node-value is N, a split-value is S and a pointer is R.

In general, a node of a G-tree will be arranged as in Fig. 6.

It is assumed that each of these nodes will be stored on a page of given page size, P. Thus if a node contains n node-values, $k_1, \dots, k_n$, $m+1$ references $r_0, r_1, \dots, r_m$ and m split-values, $s_1, \dots, s_m$, a necessary constraint is that

$$nN + (m+1)R + mS \leq P.$$

We will assume that the G-trees are ordered, i.e. that if $K(r_i)$ denotes the set of node-values contained in the tree pointed to by r_i, then

$$(\forall k \in K(r_0))(k \leq s_1),$$
$$(\forall k \in K(r_i))(s_i < k \leq s_{i+1}), \quad i = 1, \dots, m-1,$$
$$(\forall k \in K(r_m))(s_m < k).$$

The major cost in retrieving a key from a G-tree is given by the number of page accesses. Thus, if p_j denotes the probability that a key k_f occurs as a search argument, then the expected cost of a successful search is given by

$$\sum p_j \,(\text{level } k_j + 1)$$

where the sum is taken over all the node-values in the G-tree. An optimal G-tree is one designed to minimize this value. Algorithms are required to construct (near) optimal G-trees for given data.

Dynamic G-trees may be useful in organising data where a small proportion is commonly accessed. One solution is to insist that, with each reference in a node, is associated one or more node-values. These node-values will represent nodes which could appear in the subtree pointed to by the reference but currently have a greater frequency than any node so doing. It is then always possible to interchange a node-value whose frequency increases with some node-value in

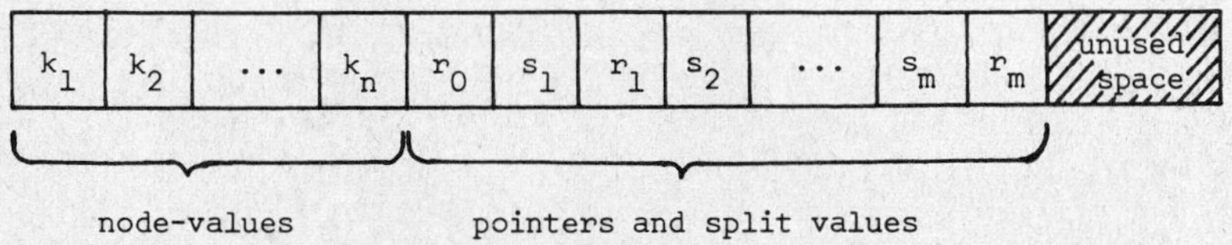

Fig. 6.

the parent. Similarly, any node-value whose frequency decreases can be simply interchanged with a node-value appearing at the root of one of its immediate descendents. Such interchanging can be automated so that commonly accessed node-values filter their way towards the root of the tree. Insisting that every node has a minimum number of node-values is probably unwise in practical circumstances and thus making a G-tree dynamic is not a trivial problem.

References

[1] R. Bayer and E. McCreight, "Organization and maintenance of large ordered indexes", *Acta Informatica* 1 (1972) 173–189.

[2] J. Bruno and E.G. Coffman, "Nearly optimal binary search trees", *Proceedings of IFIP Conference, Ljublyana, Yugoslavia* (North-Holland, Amsterdam, 1972).

[3] D. Comer, "The ubiquitous B-tree, *Computing Surveys* 11 (1979) 121–137.

[4] M.L. Fredman, "Two applications of a probabilistic search technique: sorting $x+y$ and building balanced search trees", *Proceedings of the 7th Association for Computing Machinery Symposium on the Theory of Computing*, Albuquerque (1975).

[5] H.F. Gaines, *Cryptanalysis* (Dover, New York, 1956).

[6] T.C. Hu, "Some results and problems in binary trees", in: R. Rustin, ed., *Combinatorial algorithms* (Academic Press, New York, 1972) pp. 11–15.

[7] T.C. Hu and A.C. Tucker, "Optimal computer search trees and variable-length alphabetic codes", *Journal of the Society of Industrial and Applied Mathematics on Applied Mathematics* (1971) 514–532.

[8] D.E. Knuth, "Optimum binary search trees", *Acta Informatica* 1 (1971) 14–25.

[9] D.E. Knuth, *The art of computer programming. Volume 1: fundamental algorithms* (Addison-Wesley, Reading, MA, 1968).

[10] D.E. Knuth, *The art of computer programming. Volume 3: sorting and searching* (Addison-Wesley, Reading, MA, 1973).

[11] K. Melhorn, "Nearly optimal binary search trees", *Acta Informatica* 5 (1975) 287–295.

[12] B.A. Shiel, "Median Split trees: a fast look-up technique for frequently occurring keys", *Communications of the Association for Computing Machinery* 11 (1978) 947–958.

Mathematical Programming Study 13 (1980) 121–134.
North-Holland Publishing Company

HEURISTIC ANALYSIS, LINEAR PROGRAMMING AND BRANCH AND BOUND

Laurence A. WOLSEY*

London School of Economics, London, Great Britain

Received 1 February 1980

"*The methods used for designing such (heuristic) algorithms tend to be rather problem specific, although a few guiding principles have been identified and can provide a useful starting point*".

M.R. Garey and D.S. Johnson: Computers and Intractibility [11, Ch. 6, p. 122].

We consider two questions arising in the analysis of heuristic algorithms.

(i) Is there a general procedure involved when analysing a particular problem heuristic?

(ii) How can heuristic procedures be incorporated into optimising algorithms such as branch and bound?

In answer to (i) we present one possible procedure, and discuss the cutting stock and travelling salesman problems from this point of view. Noting that the analysis of a heuristic is often based on a linear programming relaxation, we then show how certain heuristics can be integrated into enumeration schemes to produce branch and bound algorithms whose worst case behaviour steadily improves as the enumeration develops. We take the multidimensional knapsack problem, the uncapacitated K-location problem, and the travelling salesman problem as examples.

Key words: Algorithm Analysis, Benders' Algorithm, Bin Packing, Branch and Bound, Duality Gaps, Dynamic Programming, (Euclidean) Travelling Salesman, Heuristic, Longest Hamiltonian Tour, Matching Heuristic, (Minimum Length) Eulerian Tours, (Multidimensional) Knapsack, Optimising Problems, Uncapacitated k-location.

1. Introduction

Many people would agree that the quotation above would be apt with "designing" replaced by "analysing". The aim of this paper is to look for one or two guiding principles, and in particular principles relating the analysis of heuristics to such traditional preoccupations of operations researchers as linear programming and branch and bound.

In Section 2 we suggest a procedure for obtaining worst case results of the form: $Z^{H} \leq rZ + s$ for a given (minimising) combinatorial optimization problem, where Z is the optimal value, Z^{H} is the heuristic value, and $r \geq 1$. We assume the

* This research was supported by a Senior Visiting Research Fellowship from the Science Research Council, while the author was on leave from CORE, Université Catholique de Louvain at Louvain-la-Neuve, Belgium.

problem can be formulated as a linear integer program, and the essential step is to relate the heuristic solution to a dual feasible solution of the given integer problem. As the dual feasible solutions are often linear, the results obtained are often of the stronger form: $Z^{\mathrm{H}} \leq rZ^{\mathrm{LP}} + s$, where Z^{LP} is the optimal value of the linear programming relaxation of the problem. Such cases in turn imply results about duality gaps: $Z \leq rZ^{\mathrm{LP}} + s$ for the given class of problems. As examples we examine the first fit heuristic for the cutting stock problem, and heuristics for two versions of the optimal Hamiltonian tour problem.

One feature common to various heuristics is the use of partial enumeration. In Section 3 we attempt to integrate heuristics giving bounds of the form: $Z^{\mathrm{H}} \leq rZ^{\mathrm{LP}}$ with partial enumeration to obtain branch and bound algorithms that use both the heuristic and linear programming bounds systematically. For instance if R is the level of enumeration, Z_R^{LP} denotes the problem lower bound, obtained as the smallest linear programming bound over the active nodes, and Z_R^{H} is the value of the best heuristic solution found, one would like results of the form: $Z_R^{\mathrm{H}} \leq Z_R^{\mathrm{LP}}(r - \gamma(R))$ where $\gamma(R)$ increases strictly with R. As examples demonstrating such behaviour we take the multi-dimensional knapsack problem, and the uncapacitated K-plant location problem. Then to indicate some of the difficulties we look at the Euclidean travelling salesman problem.

2. Worst case heuristic analysis

We consider what is involved in proving a worst case result for a combinatorial optimisation problem described as a linear integer program:

$$\text{(P)} \qquad \begin{aligned} Z = \min\ & cx, \\ & Ax \geq b, \\ & x \geq 0 \text{ and integer,} \end{aligned}$$

where A, b have integer coefficients, $A(m \times n)$, and its linear programming relaxation is $Z^{\mathrm{LP}} = \min\{cx\colon Ax \geq b, x \geq 0\}$.

Let Z^{H} be the value of a heuristic solution to (P). Suppose now that one can find a function $F : Z^m \to R$ with the following properties:

(i) F is subadditive and nondecreasing i.e.

$$F(u) + F(v) \geq F(u + v), \quad u, v \in Z^m,$$

$$F(u) \leq F(v) \quad \text{if } u_i \leq v_i,\ i = 1, \dots, m,$$

(ii) $F(a_j) \leq c_j, \quad j = 1, \dots, n,$

(iii) $F(b) \geq (Z^{\mathrm{H}} - s)/r$ where $r \geq 1$.

Then we obtain the following result:

Theorem 1. $Z^{\mathrm{H}} \leq rZ + s$.

Proof. Let x^* be an optimal solution to (P). Then

$$Z = \sum_{j=1}^{n} c_j x_j^* \geq \sum_{j=1}^{n} F(a_j) x_j^* \geq F\left(\sum_{j=1}^{n} a_j x_j^*\right) \geq F(b) \geq (Z^{\mathrm{H}} - s)/r$$

where the first inequality follows from (ii), the second and third from (i), and the last from (iii).

Restating the properties in the terminology of integer programming duality theory, see [13, 17], we write the dual of (P) as:

$$W = \max F(b),$$

(D) $$F(a_j) \leq c_j, \quad j = 1, \dots, n,$$

F subadditive and nondecreasing

and observe that the above proof is just a proof of the Weak Duality Theorem:

"*If F is dual feasible, $F(b)$ is a lower bound on the value of Z*".

This suggests the following description of heuristic analysis:

Ideally given the heuristic solution one is required to find a relationship between the value of the heuristic solution Z^{H} and the optimal value Z. In practice Z is unknown so one is forced to use some lower bound on Z. One general procedure for finding lower bounds on Z is to solve relaxations of (P), or more generally to look for feasible solutions to its dual (D). (Note that if (P′) is a relaxation of (P), its dual (D′) is a restriction of (D).)

In practice finding subadditive functions is not simple, and as one often knows much more about the linear programming relaxation of (P), one often only finds linear solutions to (D) of the form:

$$F(d) = \sum_{i=1}^{m} u_i d_i, \quad \text{with } u \in R_+^m, \quad d \in R^m.$$

In such cases one immediately obtains a slightly stronger result:

Theorem 2. *If $F(d) = ud$, $u \geq 0$ is feasible in* (D), *and $ub \geq (Z^{\mathrm{H}} - s)/r$, then $Z^{\mathrm{H}} \leq rZ^{\mathrm{LP}} + s$.*

More generally if F is feasible for the dual of a relaxation (A) of (P) with value Z^{A}, we obtain $Z^{\mathrm{H}} \leq rZ^{\mathrm{A}} + s$.

Example 1 [12, 14]. *The bin packing, or cutting stock problem.* Gilmore and Gomory gave the formulation:

$$Z = \min 1 \cdot x,$$

$$\sum_j a_j x_j \geq b,$$

$x \geq 0$ and integer

where b_i is the number of pieces of length l_i to be packed (cut), a_j is bin packing (cutting pattern) j, and each bin has unit length.

The heuristic "First Fit" takes the pieces in any order and puts each piece into the first bin that still has room for it. Let Z^{FF} be the number of bins required.

Theorem 3. $Z^{\mathrm{FF}} \leq 1.7 Z^{\mathrm{LP}} + 2$.

Proof. In the classic paper on bin packing [14] Johnson et al. construct a function W with the property that $F(a_j) = \sum_i W(l_i) a_{ij} \leq 1 \, \forall j$, and F is subadditive and nondecreasing. As

$$F(b) \geq \frac{Z^{\mathrm{FF}} - 2}{1.7},$$

they conclude that $Z^{\mathrm{FF}} \leq 1.7 Z + 2$. As the function

$$F(d) = \sum_{i=1}^{m} W(l_i) d_i$$

is actually linear, the strengthened result follows immediately by Theorem 2.

Example 2 [7]. *Longest undirected hamiltonian tours.* Given a complete graph $G = (N, E)$, with edges of length $c_e \geq 0$, $e \in E$, this has the standard formulation:

$$\text{(P)} \qquad \begin{aligned} Z = \max & \sum_{e \in E} c_e x_e, \\ \text{s.t.} & \sum_{e \in V_i} x_e = 2, \quad i \in N, \\ & \sum_{e \in S} x_e \leq |S| - 1, \quad \phi \subset S \subset N, \\ & x_e \in \{0, 1\}, \quad e \in E \end{aligned}$$

where V_i = the set of edges incident with vertex $i \in N$, and $e \in S$ only if both endpoints of edge e lie in S.

Relaxing (P) we obtain the problem:

$$\begin{aligned} Z^{\mathrm{A}} = \max & \sum_{e \in E} c_e x_e, \\ & \sum_{e \in V_i} x_e = 2, \quad i \in N, \\ & x_e \geq 0, \quad e \in E \end{aligned}$$

which is solvable as an assignment problem, with dual:

$$\min\left\{2 \sum_{i=1}^{n} u_i, \quad u_i + u_j \geq c_e, \quad e = (i, j) \in E\right\}.$$

As heuristic we consider the best neighbour heuristic which specifies an initial vertex and chooses an edge of maximum weight adjacent to it. It then continues from the node just reached choosing the best edge subject to the collection still being part of a tour. Let Z^{N} be its length.

Theorem 4 [7]. $Z^{\mathrm{N}} \geq \frac{1}{2} Z^{\mathrm{A}}$.

Proof. Suppose the edges selected in order are $e_{12}, e_{23}, \ldots, e_{n1}$. Let $u_i = c_{i,i+1}$, $i = 1, \ldots, n-1$, $u_n = c_{1n}$. u is dual feasible for (A) as $i < j$ implies $u_i \geq c_{ij}$, and hence

$$Z \leq Z^{\mathrm{A}} \leq F(b) = 2 \sum_{i=1}^{n} u_i = 2Z^{\mathrm{N}}.$$

Example 3. *Minimum length Eulerian tours*: Given a graph $G = (N, E)$ with $c_e \geq 0$, this can be formulated as:

$$Z = \min \sum_{e \in E} c_e x_e,$$

$$\text{(P)} \qquad \sum_{e \in V_i} x_e = 2 + 2w_i, \quad \forall i \in N,$$

$$\sum_{e \in (S, \bar{S})} x_e \geq 2, \quad \forall \phi \subset S \subset N,$$

$$x_e \geq 0 \text{ and integer } e \in E, \quad w_i \geq 0 \text{ and integer } i \in N$$

where $e \in (S, \bar{S})$ implies that one endpoint of e lies in S, and the other in $\bar{S} = N \setminus S$.

We shall show that two well-known heuristics for the Euclidean travelling salesman problem, the "tree" heuristic, with value Z^{T}, and the "Christofides" heuristic with value Z^{C} actually apply to this problem, and lead to:

Theorem 5. $Z^{\mathrm{T}} \leq 2Z^{\mathrm{LP}}$, $Z^{\mathrm{C}} \leq \frac{3}{2} Z^{\mathrm{LP}}$.

The "tree" heuristic [16] involves finding a minimum cost spanning tree T in G, and then duplicating each edge of the tree.

The "Christofides" heuristic [4] involves first finding a minimum cost spanning tree T, and then finding the minimum length Eulerian tour containing T.

Proof of Theorem 5. Let $c(T)$ denote the length of the minimum cost spanning tree, and $c(E')$ be the length of the set of edges E' added in the Christofides heuristic. We show below that $c(T) \leq Z^{\mathrm{LP}}$ and $c(E') \leq \frac{1}{2} Z^{\mathrm{LP}}$. Then as $Z^{\mathrm{T}} = 2c(T)$ and $Z^{\mathrm{C}} = c(T) + c(E')$ the result follows.

Note that $\sum_{e \in V_i} x_e \geq 2 \ \forall i \in N$ implies that:

$$2\sum_{e\in\bar{S}} x_e + \sum_{(S,\bar{S})} x_e \geq 2|\bar{S}|.$$

Combined with $\sum_{(S,\bar{S})} x_e \geq 2$, this gives:

$$\sum_{e\in\bar{S}} x_e + \sum_{(S,\bar{S})} x_e \geq |\bar{S}| + 1.$$

Therefore

$$\min \sum_{e\in E} c_e x_e,$$

$$\sum_{e\notin S} x_e \geq n - |S|, \quad \phi \subset S \subset N,$$

$$x_e \geq 0$$

is a valid relaxation of (LP). The extreme points of this unbounded polyhedron are the spanning trees, see [9], and hence $c(T) \leq Z^{\text{LP}}$.

To find a spanning tour containing T, let σ/ϵ denote the vertices of N of odd/even degree in T. Let E' be a set for which $T \cup E'$ is a spanning tour. Suppose now that $|S \cap \sigma|$ is odd for some $S \subset N$. For $i \in S \cap \sigma$, the degree of E' at i is odd, and for $i \in S \cap \epsilon$ the degree is even. It follows that $|(S,\bar{S}) \cap E'| \geq 1$. Conversely note that $|(R,\bar{R}) \cap T| < 2$ is only possible if $|R \cap \sigma|$ is odd. We have now shown that $T \cup E'$ is a Eulerian tour if and only if E' is a feasible solution of:

$$\min \sum_{e\in E} c_e x_e,$$

$$\sum_{e\in V_i} x_e = 1 + 2w_i, \quad i \in \sigma,$$

$$\sum_{e\in V_i} x_e = 0 + 2w_i, \quad i \in \epsilon,$$

$$\sum_{e\in(S,\bar{S})} x_e \geq 1 \quad \forall |S \cap \sigma| \text{ odd},$$

$$x_e \geq 0 \text{ and integer}, \; w_i \geq 0 \text{ and integer}.$$

This problem is the Chinese postman problem, analysed in detail by Edmonds and Johnson [6]. They show that it is equivalent to:

$$Z^{E'} = \min \sum_{e\in E} c_e x_e,$$

$$\sum_{e\in(S,\bar{S})} x_e \geq 1 \quad \forall |S \cap \sigma| \text{ odd},$$

$$x_e \geq 0$$

and that its optimal solution is a set of edge disjoint shortest paths between pairs of vertices of σ.

Observing that

$$\min \sum_{E} c_e x_e,$$

$$\sum_{(S,\bar{S})} x_e \geq 2 \quad \forall |S \cap \sigma| \text{ odd}$$

$$x_e \geq 0$$

is a relaxation of (LP), we finally have $Z^{E'} \leq \frac{1}{2} Z^{\mathrm{LP}}$.

Note that when the $\{c_e\}_{e \in E}$ satisfy the triangle inequality, the minimum length Eulerian tour necessarily becomes a Hamiltonian tour, and the set of shortest edge disjoint paths E' becomes a matching on the subgraph induced by σ. More generally we note that a minimum length Eulerian tour can be found by solving the Euclidean travelling salesman problem where the edge lengths $\{c'_e\}$ are the shortest distances between nodes based on the original distance matrix $\{c_e\}$. It is also easily shown that LP is a relaxation of the linear program obtained from the standard travelling salesman problem formulation of Example 2.

Other examples of worst case analysis based on Theorems 1 or 2 can be found in [3, 5, 7, 14].

3. Embedding heuristics into optimisation algorithms

We have just seen several examples of heuristic analysis where the value of the heuristic solution is expressed in terms of a linear programming bound on the optimal value. Another feature common to certain heuristics is that they are based on partial enumeration. Here we attempt to combine these two properties to obtain implicit enumeration, or branch and bound algorithms.

The basic features we need to describe for a problem (P) are

(a) the linear programming relaxation at each node,

(b) the heuristic algorithm used at each node, and

(c) the branching procedure used.

We suppose that the enumeration tree has been developed explicitly, or implicitly down to level R. We let Z_R^{H} denote the value of the best heuristic solution found at that stage, and let Z_R^{LP} be the worst linear programming bound over the nodes active at level R.

Example 4 [2]. *The multidimensional knapsack problem.* This has the standard formulation:

$$Z = \max cx,$$
$$\text{(P)} \qquad Ax \le b,$$
$$x \ge 0 \text{ and integer}$$

where we take c, A, b to be non-negative integer, and A is $m \times n$.

Basics of the algorithm. Order the variables so that $c_1 \ge c_2 \ge \cdots \ge c_n$. Each node is defined by a non-negative integer vector y. If p is the largest nonzero coordinate of y, the subproblem at node y is:

$$Z_y = cy + \max \sum_{j=1}^{n} c_j x_j,$$
$$(\mathrm{P}_y) \qquad \sum_{j=1}^{n} a_j x_j \le b - \sum_{j=1}^{n} a_j y_j,$$
$$x_j \ge 0 \text{ and integer } j = 1, \dots, n.$$
$$x_j = 0,\ j = 1, \dots, p-1,\ x_j = 0 \text{ if } a_j \nleq b - \sum_{j=1}^{n} a_j y_j,\ j = p, \dots, n.$$

Nodes at level k are those for which $\sum_{j=1}^{n} y_j = k$. A list of nodes is maintained with nodes removed from the top, and added at the bottom. Termination occurs when the list is empty.

Let (LP_y) denote the linear programming relaxation of (P_y) with basic solution x^*, and value $Z_y^{\mathrm{LP}} = c(y + x^*)$.

The *upper bound* at node y is taken to be Z_y^{LP}.

The *lower bound* is obtained by rounding down x^* to the nearest integer $\lfloor x^* \rfloor$, so that $Z_y^{\mathrm{H}} = c(y + \lfloor x^* \rfloor)$.

To *branch* from node y, add the nodes $\{y^t\}_{t=p}^{n}$ where $y^t = y + e_t$, unless $A(y + e_t) \nleq b$, where e_i denotes the ith unit vector.

The *fathoming and updating rules* are standard.

First we observe:

Proposition 1. $Z_y^{\mathrm{LP}} - Z_y^{\mathrm{H}} < mc_p$.

Proof.

$$Z_y^{\mathrm{LP}} - Z_y^{\mathrm{H}} = c(y + x^*) - c(y + \lfloor x^* \rfloor) = c(x^* - \lfloor x^* \rfloor) < m \max\{c_i\}_{i=p}^{n} = mc_p,$$

as x^* is basic.

This combined with the ordering of the variables and the branching rule leads to:

Theorem 5. *Either the algorithm stops with an optimal solution before level R is terminated, or*

$$Z_R^{\mathrm{H}} > \left(1 - \frac{m}{R}\right) Z_R^{\mathrm{LP}}$$

where

$$Z_R^{\mathrm{H}} = \max_{\Sigma y_i \le R} Z_y^{\mathrm{H}}, \quad \text{and} \quad Z_R^{\mathrm{LP}} = \max_{\substack{\Sigma y_j = R \\ y \text{ active}}} Z_y^{\mathrm{LP}}.$$

Proof. Suppose the algorithm has not terminated when all nodes on the list with $\sum_{j=1}^n y_j = R$ have been removed. This implies that $Z_R^{\mathrm{LP}} = Z_y^{\mathrm{LP}}$ for some y^* with $\sum_{j=1}^n y_j^* = R$. Let q be the last nonzero coordinate of y^*. Then $Z_R^{\mathrm{H}} \ge Z_{y^*}^{\mathrm{H}}$ by definition of Z_R^{H}, and $Z_{y^*}^{\mathrm{H}} \ge Z_{y^*}^{\mathrm{LP}} - mc_q$ by Proposition 1. Also $Z_{y^*}^{\mathrm{LP}} \ge cy^* \ge Rc_q$, and hence

$$\frac{Z_R^{\mathrm{LP}} - Z_R^{\mathrm{H}}}{Z_R^{\mathrm{LP}}} \le \frac{Z_{y^*}^{\mathrm{LP}} - Z_{y^*}^{\mathrm{H}}}{Z_{y^*}^{\mathrm{LP}}} \le \frac{mc_q}{Rc_q} = \frac{m}{R}.$$

Note that the theorem and its proof are essentially due to Chandra et al. [2] and only the interpretation is new.

Example 5 [5, 15]. *The simple K-plant location problem.* Letting $I = \{1, \dots, m\}$, $N = \{1, \dots, n\}$, this can be formulated as: $(c_{ij} \ge 0)$.

$$\begin{aligned} Z = \max \sum_{i \in I} \sum_{j \in N} & c_{ij} x_{ij}, \\ \sum_{j \in N} x_{ij} &\le 1 \quad \forall i \in I, \\ x_{ij} &\le y_j \quad \forall i \in I, j \in N, \\ \sum_{j \in N} y_j &= K, \\ x_{ij} &\ge 0 \quad i \in I, j \in N, \qquad y_j \in \{0, 1\}, \quad j \in N. \end{aligned}$$

Notation. For $\phi \ne S \subseteq N$, let

$$z(S) = \sum_{i=1}^m \max_{j \in S} c_{ij}$$

be the value of the above program when $y = y^S$ the characteristic vector of S. Thus $z(S)$ is the value of opening plants at locations S, and then assigning customers $i = 1, \dots, m$ optimally.

Basics of the algorithm. Each node is defined by a set $T \subseteq N$ of the plants that are open, and a set $U \subseteq N - T$ of plants that are closed. The plants in $Q = N - (T \cup U)$ are still free. The subproblem at node T, U is

$$
(\mathrm{P}_{T,U}) \qquad
\begin{aligned}
Z_{T,U} &= z(T) + \sum_{i\in I}\sum_{j\in Q}(c_{ij}-u_i)x_{ij},\\
&\sum_{j\in Q} x_{ij} \le 1 \quad \forall i\in I,\\
&x_{ij}\le y_j \quad \forall i\in I,\ \ j\in Q,\\
&\sum_{j\in N} y_j = K-|T|.\\
&x_{ij}\ge 0,\quad i\in I,\ \ j\in Q, \qquad y_j\in\{0,1\},\ \ j\in Q
\end{aligned}
$$

where $u_i = \max_{j\in T} c_{ij}$.

Nodes at level k are those for which $|T| = k$.

The *upper bound* $Z^{\mathrm{LP}}_{T,U}$ at node (T, U) is obtained from the linear programming relaxation of $(\mathrm{P}_{T,U})$.

The *lower bound* $Z^{\mathrm{H}}_{T,U}$ is obtained by applying the greedy heuristic to $(\mathrm{P}_{T,U})$, see [4]. Starting with $S^0 = T$ choose $S^1, S^2, \dots, S^{K-k}$ by setting $S^t = S^{t-1}\cup\{j^*\}$ where

$$z(S^{t-1}\cup\{j^*\}) = \max_{j\in N-S^{t-1}-U} z(S^{t-1}\cup\{j\}).$$

S^{K-k} is the heuristic solution, and $Z^{\mathrm{H}}_{T,U} = z(S^{K-k})$.

To *branch* from node (T, U), order the variables of $Q = \{j_1, \dots, j_r\}$ so that $z(T\cup\{j_1\}) \ge \cdots \ge z(T\cup\{j_r\})$, and add the nodes $(T^t, U^t)_{t=1}^r$ where $T^t = T\cup\{j_t\}$, $U^t = U\cup\{j_1, \dots, j_{t-1}\}$.

The *fathoming and updating rules* are standard.

Proposition 2. $(Z^{\mathrm{H}}_{T,U} - z(T)) \ge (Z^{\mathrm{LP}}_{T,U} - z(T))\left(1-\dfrac{1}{e}\right)$.

Theorem 6. *Either the algorithm stops with an optimal solution before level R is terminated, or*

$$Z^{\mathrm{H}}_R \ge \left(1-\frac{1}{e}\left(1-\frac{R}{K}\right)\right) Z^{\mathrm{LP}}_R.$$

These results are proved in [15], as a special case of the problem: $\max_{S\subseteq N}\{z(S)\colon |S|\le K\}$ where z is a submodular, nondecreasing function. There it is shown that a version of Theorem 6 holds for this more general problem class, and in particular for the capacitated K-location problem.

The above results depend both on having an appropriate enumeration procedure, and a lower/upper bound relationship which is valid for the incremental problem value (i.e. $Z_y - cy$, $Z_{T,U} - z(T)$) at each node. In re-examining the two heuristics for the Eulerian tour problem we shall see below how first one and then both properties fail.

Example 6. *The Euclidean travelling salesman problem.* This has the same formulation as Example 2, except for objective min $\sum_E c_w x_e$, with $\{c_e\}_{e\in E}$ satisfying the triangle inequality. Alternatively it is a special case of Example 3.

We first adapt the "tree" heuristic analysed earlier.

Basics of the algorithm. Each node is defined by a simple path Q of edges $e_1, \dots, e_t$, denoting all tours containing that path. Nodes at level k are paths containing k edges.

$$(\mathrm{P}_Q) \qquad Z_Q = \min\Big\{\sum_{e\in E} c_e x_e : \sum_{e\in V_1} x_e = 2\,\forall i\in N, \sum_{e\in S} x_e \le |S|-1\,\forall_\phi \subset S\subset N,$$

$$x_e = 1, e\in Q, x_e \in \{0,1\} \text{ otherwise}\Big\}.$$

The *lower bound* Z_Q^{LP} at Q is obtained from the linear programming relaxation.

The *upper bound* at Q is found as follows. Let L be the vertices of N that are either disjoint from the path Q or endpoints $\{s, t\}$ of Q. Find a minimum spanning tree T^* on the subgraph induced by L. Repeat each edge of T^* twice, and using the triangle inequality convert into a Hamiltonian path through L from s to t whose length $\le 2c(T^*)$. Let Z_Q^{T} be the length of the resulting tour containing Q.

Proposition 3. $Z_Q^{\mathrm{T}} - c(Q) \le 2(Z_Q^{\mathrm{LP}} - c(Q))$.

Proof. Showing that $c(T^*) \le Z_Q^{\mathrm{LP}} - c(Q)$ is similar to the argument used in the proof of Theorem 5. It remains to show that a Hamiltonian path of length $\le 2c(T^*)$ can be constructed. Let the $s-t$ path in T^* be $s = j_0, j_1, \dots, j_r$. T^* then consists of this path plus trees $T_0, T_1, \dots, T_r$ rooted at $j_0, \dots, j_r$. Based on the triangle inequality, the required path is obtained with $Z_Q^{\mathrm{H}} - c(Q) \le 2c(T^*)$ provided that starting from $j_0 = s$, all vertices of T_{i-1} are visited before moving to j_i, and at the last step all vertices of T_r are visited before finishing with its root $j_r = t$.

To obtain a branch and bound result, we apparently need a *branching* procedure with the property that if Q is a node at level k $c(Q) \ge (k/n)Z_Q^{\mathrm{LP}}$. Failing this, we know that the optimal tour for (P) necessarily contains some simple path Q' with $|Q'| = k$, such that $c(Q') \ge (k/n)Z_{Q'}$. This gives:

Theorem 7. $Z_R^{\mathrm{T}} \le Z(2-(R/n))$.

Proof. Let Q' be as above with $|Q'| = R$. Then

$$Z_R^{\mathrm{T}} \le Z_{Q'}^{\mathrm{T}} \le 2Z_{Q'} - c(Q') \le Z\Big(2 - \frac{R}{n}\Big)$$

as $Z_{Q'} = Z$.

One would like to carry out a similar analysis by adopting the Christofides' heuristic, so as to take the tree T^*, and then find a perfect matching M on σ where $\sigma = \{$vertices of $L - \{s, t\}$ having odd degree in T^*, plus s and t if either has even degree in $T^*\}$. If Z_Q^{C} is the length of this heuristic tour, we obtain $Z_Q^{\mathrm{C}} \le \frac{3}{2} Z_Q$ which appears insufficient for the construction of an enumerative heuristic.

However, by sharpening this analysis, McDiarmid has shown:

Theorem 8. $\min_{|Q|=R} \min\{Z_Q^{\mathrm{T}}, Z_Q^{\mathrm{C}}\} \le Z(\frac{5}{3} - \frac{2}{3}(R/n))$.

Proof. The proof of Proposition 3 shows in fact that $Z_Q^{\mathrm{H}} - c(Q) \le 2(Z_Q^{\mathrm{LP}} - c(Q)) - c_{s,t}$. For the Christofides heuristic one obtains $c(M) \le \frac{1}{2}(Z_Q^{\mathrm{LP}} - c(Q) + c_{s,t})$, and hence $Z_Q^{\mathrm{C}} - c(Q) \le \frac{3}{2}(Z_Q^{\mathrm{LP}} - c(Q)) + \frac{1}{2} c_{s,t}$. Hence

$$\begin{aligned}\min\{Z_Q^{\mathrm{T}}, Z_Q^{\mathrm{C}}\} - c(Q) &\le \min[\tfrac{3}{2}(Z_Q^{\mathrm{LP}} - c(Q)) + \tfrac{1}{2} c_{s,t}, 2(Z_Q^{\mathrm{LP}} - c(Q)) - c_{s,t}] \\ &\le \tfrac{5}{3}(Z_Q^{\mathrm{LP}} - c(Q))\end{aligned}$$

where the last inequality comes from equating the two terms. The proof now follows that of Theorem 7.

Frieze [8] has obtained a similar enumerative result of the form $Z^{\mathrm{H}} \ge Z(\frac{2}{3} + \frac{1}{3}(R/n) - (1/n))$ for the longest Hamiltonian tour problem using the matching heuristic proposed in [7].

4. Further observations

An alternative algorithm that has been proposed for certain mixed integer programming problems is that due to Benders [1]. It is somewhat surprising that both the positive results on enumerative heuristics, Theorems 5 and 6, lead to similar results relating the heuristic value to a "Benders-like" relaxation of the problem (P). For the multi-dimensional knapsack problem this is of little interest, but the result for the uncapacitated K-plant location problem is a result about Benders' algorithm, and is given in detail in [12].

As in Section 2 the branch and bound results of the previous section can be expressed in terms of solutions to the dual of (P). However, these dual solutions are not subadditive, and satisfy the more general dual (D′):

$$\begin{aligned}\max\{F(b)\colon\ & F(Ax) \le cx\ \forall x \ge 0 \text{ and integer}, \\ & F: R^m \to R \text{ and nondecreasing}\},\end{aligned}$$

see [17].

The tie we have suggested between enumerative heuristics and branch and bound is somewhat reminiscent of the equivalence [10] between "fully polynomial approximation schemes", and "dynamic programming", and raises the question whether there is not a more precise relationship.

Throughout this paper we have been re-examining heuristics that have already been analysed. Even so, many questions remain. Is the $\frac{11}{9}$ bound for bin-packing using the first-fit decreasing heuristic [14] valid for the linear programming relaxation, and what is the worst value of the duality gap for this problem? For the Euclidean travelling salesman problem we have shown that the duality gap ratio never exceeds $\frac{3}{2}$. The largest gap we are aware of is $\frac{8}{7}$. Is either of these values tight?

More generally we hope that we have suggested a useful heuristic for analysing new as yet undreamt of heuristics.

Acknowledgment

We are grateful to Colin McDiarmid for his comments and suggestions, and in particular for Theorem 8.

References

[1] J.F. Benders, "Partitioning procedures for solving mixed-variables programming problems", *Numerische Mathematik* 4 (1962) 238–252.

[2] A.K. Chandra, D.S. Hirchberg and C.K. Wong, "Approximate algorithms for some generalised knapsack problems", *Theoretical Computer Science* 3 (1976) 293–304.

[3] V. Chvatal, "The covering problem", in: *Lecture notes on heuristics* (McGill University, 1978).

[4] N. Christofides, "Worst case analysis of a new heuristic for the travelling salesman problem", GSIA report No. 388, Carnegie-Mellon University (1976).

[5] G. Cornuejols, M.L. Fisher and G.L. Nemhauser, "Location of bank accounts to optimize float: an analytic study of exact and approximate algorithms", *Management Science* 23 (1977) 789–810.

[6] J. Edmonds and E.L. Johnson, "Matching, Euler tours and the chinese postman", *Mathematical Programming* 5 (1973) 88–124.

[7] M.L. Fisher, G.L. Nemhauser and L.A. Wolsey, "An analysis of approximations for finding a maximum weight Hamiltonian circuit", *Operations Research* 27 (1979) 799–809.

[8] A.M. Frieze, "Worst case analysis of algorithms for travelling salesman problems", Technical report, Department of Computer Science and Statistics, Queen Mary College, London (1978).

[9] D.R. Fulkerson, "Blocking and anti-blocking pairs of polyhedra", *Mathematical Programming* 1 (1971) 168–194.

[10] M.R.Garey and D.S. Johnson, "Strong NP-completeness results: motivations, examples and implications", *Journal of the Association of Computing Machinery* 25 (1978) 499–508.

[11] M.R. Garey and D.S. Johnson, *Computers and intractibility* (W.H. Freeman, San Francisco, CA, 1979).

[12] P.C. Gilmore and R.E. Gomory, "A linear programming approach to the cutting stock problem", *Operations Research* 9 (1961) 849–859.

[13] R.G. Jeroslow, "Cutting plane theory: algebraic methods", *Discrete Mathematics* 23 (1978) 121–150.

[14] D.S. Johnson, A. Demers, J.D. Ullman, M.R. Garey and R.L. Graham, "Worst case performance bounds for simple one-dimensional packing algorithms", *Society for Industrial and Applied Mathematics Journal on Computing* 3 (1974) 299–325.

[15] G.L. Nemhauser and L.A. Wolsey, "Maximizing submodular set functions: formulations, algorithms and applications", CORE D.P 7832, University of Louvain-la-Neuve, Belgium (1978).

[16] D.J. Rosenkrantz, R.E. Stearns and P.M. Lewis, "An analysis of several heuristics for the travelling salesman problem", *Society for Industrial and Applied Mathematics Journal on Computing* 6 (1977) 563–581.

[17] L.A. Wolsey, "Integer programming duality: price functions and sensitivity analysis", Mimeo, London School of Economics (1978).

Mathematical Programming Study 13 (1980) 135–142.
North-Holland Publishing Company

HEURISTIC IMPROVEMENT METHODS: HOW SHOULD STARTING SOLUTIONS BE CHOSEN?

C.J. PURSGLOVE and T.B. BOFFEY
University of Liverpool, Liverpool, Great Britain

Received 1 February 1980

A theoretical framework for improvement heuristic methods, as applied to discrete optimization problems, is put forward. The problem of choosing a suitable set of start points, from which to perform hill-climbs, is identified and some possible solutions considered. Numerical results are presented.

Key words: Combinatorics, Heuristic, Hill Climbing, Improvement, Optimization.

1. Introduction

Combinatorial problems are often remarkably difficult to solve even when the statement of the problem is quite straightforward. Classical examples of this situation are provided by the Travelling Salesman Problem and the even more difficult Quadratic Assignment Problem, both of which belong to the notorious class of NP-complete problems [1, 2]. In the case of the Travelling Salesman Problem, problems on 100 or more vertices were not amenable to exact solution until recently and problems of 3 or so times this size are still not tractable. Thus, if a solution is required for a larger problem or computing resources are restricted, a heuristic method must be used. A class of much used methods is based on starting with a feasible solution and successively making "small" adjustments which result in an improvement in the value of the objective. The "λ-opt" method of Shen Lin is a well-known example. This paper will concentrate on aspects of improvement methods as applied to combinatorial problems formulated in terms of 0–1 variables.

2. Basic concepts

Let P be a problem whose feasible solutions form a finite set F, and with objective function φ which we assume is to be minimized (since maximization problems can be trivially converted to minimization ones).

Definition 2.1. A function $N: F \to 2^F$, which associates a subset Nx with each $x \in F$, is a *neighbourhood function* (over F) if

(i) $|Nx| \geq 1$ all $x \in F$;
(ii) $x \notin Nx$ all $x \in F$.

Nx will be called the neighbourhood of x and $y \in F$ is a *neighbour* of $x \in F$ if $y \in Nx$. Clearly (F, N) is a directed graph which we will denote by NG(P) and call a *neighbourhood graph* over F (see Fig. 1 for example). Most methods implicitly restrict themselves to a subgraph SG(P) of NG(P):

Definition 2.2. The *search graph*, SG(P), for problem P is that graph (F, Γ) for which

$$\Gamma x = \{y \mid y \in Nx \text{ and } \varphi(y) < \varphi(x)\}.$$

Definition 2.3. A solution $x \in F$ is *locally optimal* (for a minimization problem) with respect to neighbourhood function N if $\varphi(x) \leq \varphi(y)$ all $y \in Nx$; that is, x is a sink of SG(P). A solution $x^* \in F$ is *globally optimal* (for a minimization problem) if $\varphi(x^*) \leq \varphi(x)$ for all $x \in F$.

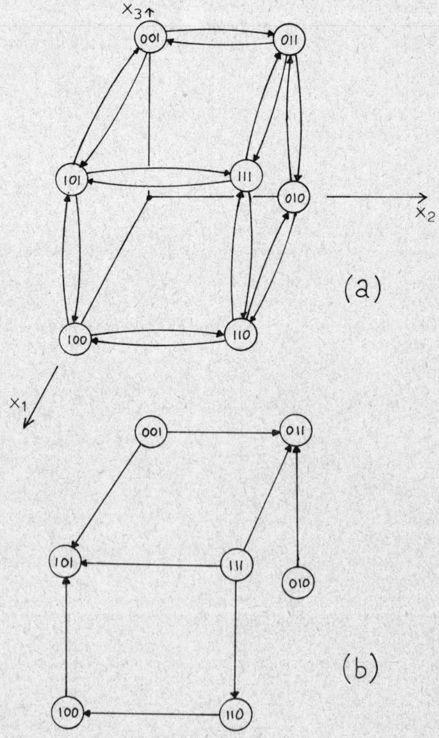

Fig. 1. Examples of NG(P) and SG(P) for some problem P.

Definition 2.4. Given a *start solution* $s \in F$, a *hill-climb* (from s) is a sequence $s = x_0, x_1, x_2, \ldots, x_t = \tilde{x}$ of elements of F such that

(i) $x_i \in \Gamma x_{i-1}$ in SG(P), $i = 1, 2, \ldots, t$.

(ii) $x_t = \tilde{x}$ is a local optimum.

Notice that, since we are developing the theory for minimization problems the hill-climbs will be "climbs down" rather than "climbs up".

Typically start points are chosen at random from F. However, it seems reasonable that this might be improved for small sets, S, of start points since

(i) the points of S may be "unevenly spread" over F since $|S|$ is small, thus leaving parts of F uncovered, and

(ii) as the calculation proceeds there is additional information provided by earlier hill-climbs that is potentially usable; unused knowledge might be expected to lead to a decrease (on average) in the quality of the best solution found.

Before proceeding further it is necessary to define precisely what we mean by distance in SG(P).

Definition 2.5. The *one-way* distance $\delta(x, y)$ and the *hamming distance* $d(x, y)$ are given by

$$\delta(x, y) = \sum_i \max(y_i - x_i, 0),$$

$$d(x, y) = \delta(x, y) + \delta(y, x),$$

for all pairs of binary vectors x, y with n components.

Two families of neighbourhood functions $\{N_{a,b}\}$ and $\{N_r\}$ have been defined by Roth [6].

$$N_{a,b}(x) = \{y \mid \delta(x, y) \leq a \ \& \ \delta(y, x) \leq b \ \& \ d(x, y) \neq 0\},$$

$$N_r(x) = \{y \mid 1 \leq d(x, y) \leq r\}.$$

Clearly $N_{a,b}(x) \subset N_{a+b}(x)$. Roth proves several results concerning these neighbourhood functions.

We can now introduce conditions which the start set S might be desired to satisfy (though not simultaneously).

3. Methods for generating start sets

Assume first that the number, m, of start points is determined beforehand. Then, noting point (i) of Section 2, we may require the m points to be spread evenly over F so that their disposition looks the same from each $x \in S$. That is we might require a "grid" of m points evenly spaced in

$$B_n = \{x \mid x_i = 0 \text{ or } 1, i = 1, 2, \ldots, n\}.$$

It seems desirable that no $x \in F$ should be "too far" from a nearest grid point and preferably that for some t the radius t Hamming sphere

$$H_t(x) = \{y \mid d(x, y) \leq t\}$$

should satisfy

(a) $\bigcup_{x \in S} H_t(x) = B_n$,

(b) $H_t(x) \cap H_t(y) = \phi$ all $x \neq y \in S$.

Grids as envisaged here correspond to *perfect codes* [4]. However, a standard result in the theory of error-correcting codes is that the only perfect codes are

(a) trivial codes;

(b) Hamming codes;

(c) the Golay code.

Unfortunately these are not suitable for our purposes as they correspond to grids with too large a number of grid points. Also the situation is not much improved if a limited amount of overlapping of Hamming spheres is allowed. Possible candidates are the grids corresponding to the Reed–Muller codes which are defined for n a power of 2 (cf. Table 1). For intermediate n, components of solution vectors could be "blocked" in some way (Section 4) though this latter suggestion can lead to grids with relatively poor distance properties.

It becomes clear that the search for grids with nearly perfect distance properties poses difficulties and in any case is probably not justified in the light of the use to be made of them, particularly if one notes further that a considerable number of the grid points may lie outside F if F is much smaller than B_n!

Another approach considered was the provision of a simple generation rule which would

(i) generate a set, S, of start points, all in F, which are well separated from each other,

(ii) be applicable for any combination of m and n, and

(iii) be such that the value of m could be decided dynamically [5]. Such a rule beginning with a list of p start points

$$x^{(1)} = (x_1^{(1)} x_2^{(1)} \cdots x_n^{(1)})$$
$$x^{(2)} = (x_1^{(2)} x_2^{(2)} \cdots x_n^{(2)})$$
$$\vdots$$
$$x^{(p)} = (x_1^{(p)} x_2^{(p)} \cdots x_n^{(p)})$$

generates the next, $x^{(p+1)}$, by building up bit by bit "from the left". Let $y|_r$ be the r-vector comprising the first r components of y (and in the same order). Let $d_r(x, y) = d(x|_r, y|_r)$ and $d_0(x, y) = 0$ for any pair of vectors x, y each of dimension at least r. The rule used was

Rule GR$(1, \infty)$

Step 1: Set $d_0(x^{(i)}, y) = 0$, $i = 1, \dots, p$ and perform Step 2 for $j = 1, 2, \dots, n$.

Step 2: Set $y_j = 1 - x_j^{(k)}$ where k is the largest index i for which $d_{j-1}(x^{(i)}, y)$ is minimal.

Step 3: If y is feasible ($y \in F$), then set $x^{(p+1)} = y$ and stop. Otherwise add y to list and return to Step 1.

Starting with (0000), rule GR$(1, \infty)$ produces as the first 8 start points the points of the Reed–Muller grid $R(1,2)$ shown in Table 1. Again, for $n = 8$, the points $x^{(1)}, x^{(2)}, \dots, x^{(8)}$ obtained by GR$(1, \infty)$ are all points of the Reed–Muller grid $R(1,3)$ (Table 1). However, the next 8 points $x^{(9)}, \dots, x^{(16)}$ vary somewhat from the remaining 8 points of $R(1,3)$.

Of course, since the number m is not predetermined GR$(1, \infty)$ chooses $x^{(p+1)}$ without regard to "later" points, the resulting set of m points is generally not as good as could be obtained if knowledge of m were taken into account throughout. Nonetheless, given the requirements demanded of the generation rule it was felt that GR$(1, \infty)$ is sufficiently good and that a more complicated rule is not justified.

The sets S generated by GR$(1, \infty)$ will deviate from the ideals of "perfect distance properties" but we might anticipate that the results obtained will be much the same as using grid-based start sets. However, we now have a flexible tool, GR$(1, \infty)$, and the values and/or the distribution of local optima already obtained can be taken into account.

First, we note that in performing a hill-climb $s = x_0, x_1, x_2, \dots, x_t = \bar{x}$ from s we are implicitly performing hill-climbs from $x_1, x_2, \dots, x_t$! Thus in generating the next element of S we might aim at keeping away from $x_1, \dots, x_t$ as well. This would involve an excessive time to apply the generation rule, so we have

Table 1
The Reed–Muller grids $R(1,2)$ and $R(1,3)$

$R(1,2)$	$R(1,3)$
(0 0 0 0)	(0 0 0 0 0 0 0 0)
(0 0 1 1)	(0 0 0 0 1 1 1 1)
(0 1 0 1)	(0 0 1 1 0 0 1 1)
(0 1 1 0)	(0 1 0 1 0 1 0 1)
(1 1 1 1)	(0 0 1 1 1 1 0 0)
(1 1 0 0)	(0 1 0 1 1 0 1 0)
(1 0 1 0)	(0 1 1 0 0 1 1 0)
(1 0 0 1)	(0 1 1 0 1 0 0 1)
	(1 1 1 1 1 1 1 1)
	(1 1 1 1 0 0 0 0)
	(1 1 0 0 1 1 0 0)
	(1 0 1 0 1 0 1 0)
	(1 1 0 0 0 0 1 1)
	(1 0 1 0 0 1 0 1)
	(1 0 0 1 1 0 0 1)
	(1 0 0 1 0 1 1 0)

compromised by adding only s and $\tilde{x}$ to the list of points to be kept away from. Also the distances $d_j(x^{(i)}, y)$ are replaced by $\alpha d_j(x^{(i)}, y)$ if $x^{(i)}$ was used as a start point and by $\beta d_j(x^{(i)}, y)$ if $x^{(i)}$ was produced as a local optimum. This leads to a more general generation rule GR(α, β).

4. Experimental results

In order to illustrate the ideas developed above we apply them to 10 0–1 Knapsack Problems each with 50 variables. (Note that we use Knapsack Problems for convenience only; it is not suggested that this is a practial method for solving Knapsack Problems of this size.) The simplest neighbourhood function N_1, was used. Each problem has the form

$$\max \sum_{i=1}^{50} a_i x_i,$$

$$\text{subject to } \sum_{i=1}^{50} b_i x_i \le W,$$

$$x_i \in \{0, 1\}.$$

For five problems the coefficients a_i and b_i are chosen randomly from a discrete uniform distribution over $\{0, 1, \dots, 99\}$ and W is (0.8) $\Sigma\, b_i$. The remaining problems are somewhat more difficult with a_i and b_i again chosen at random, but constrained to satisfy $\frac{1}{2} \le a_i/b_i \le 2$, and W again set at (0.8) $\Sigma\, b_i$.

For each problem a random 50 component binary vector $\boldsymbol{\pi}$ is generated and the four methods $M_1, \dots, M_4$ are tried where:

M_1 uses random selection.

M_2 uses a blocked $R(1, 3)$ grid in which the coordinates 1–7, 8–14, 15–20, 21–26, 27–32, 33–38, 39–44, 45–50 take the values of coordinates $1, 2, \dots, 8$ of $R(1, 3)$ respectively (cf. Table 1), then a transformation is applied by complementing the ith component if $\pi_i = 1$.

M_3 uses the rule GR(1, 1), with $\boldsymbol{\pi}$ as the first start point. Thus equal weights are given to start points and local optima.

M_4 uses the rule GR(1, β), with $\boldsymbol{\pi}$ as the first start point and β very large. Thus previous start points are avoided.

Note that the same randomly chosen binary vector $\boldsymbol{\pi}$ is chosen to start off methods M_3 and M_4 and to indicate which components of the blocked grid should be complemented.

Some statistics relating to best local optima (maxima in this case) are given in Table 2 for the four methods each applied using 16 start points.

Table 2
Statistics relating to the best local optimum obtained using the methods $M_1, \dots, M_4$ applied to ten Knapsack Problems

Method	M_1	M_2	M_3	M_4
Mean	98.15	98.6	98.2	97.9
S.D.	0.96	0.59	1.31	0.98
Median	98.0	98.8	98.3	97.8
Maximum	99.7	99.5	100.0	99.6
Minimum	96.6	97.7	97.0	96.5

All numbers are percentages of the optimal solution.

5. Conclusion

The aim of the present study was to suggest ways of improving the average quality of the solutions to 0–1 problems obtained by using different methods for choosing a (small) set of start points. Efficiency was not a prime consideration at this stage and the climbing strategy used was to select "the first ξ which leads to a maximal improvement", this variant of steepest descent (or ascent) being somewhat easier to program.

Results for a set of 0–1 Knapsack Problems were presented in some detail. Further experimentation with Knapsack and other problems has indicated that the situation is less clear cut than Table 2 might suggest. Although the results of experiments to date are inconclusive it does seem that methods better than random selection exist for choosing starting solutions, and further research should be carried out.

Acknowledgment

The authors wish to express their gratitude to Dr. A. Wragg for his helpful comments concerning error-correcting codes. Thanks are also due to the University of Liverpool Computing Laboratory for computing facilities.

References

[1] R.M. Karp, "On the computational complexity of problems", *Networks* 5 (1975) 45–68.
[2] E.L. Lawler, *Combinatorial optimization: networks and matroids* (Holt–Reinhart–Winston, New York, 1976).
[3] S. Lin, "A computer solution of the Travelling Salesman Problem", *Bell Systems Technical Journal* 44 (1965) 2245–2269.

[4] F.J. MacWilliams and N.J.A. Sloane, *The theory of error-correcting codes* (2 volumes) (North-Holland, Amsterdam, 1977).
[5] S. Reiter and G. Sherman, "Discrete optimizing", *Journal of the Society for Industrial and Applied Mathematics* 13 (1965) 864–889.
[6] R.H. Roth, "An approach to solving linear optimization problems", *Journal of the Association for Computing Machinery* 17 (1970) 303–313.